SpringerBriefs in Statistics

For further volumes:
http://www.springer.com/series/8921

Adrian Pizzinga

Restricted Kalman Filtering

Theory, Methods, and Application

 Springer

Adrian Pizzinga
Department of Statistics
Institute of Mathematics and Statistics
Fluminense Federal University
Rio de Janeiro, Brazil

ISSN 2191-544X   ISSN 2191-5458 (electronic)
ISBN 978-1-4614-4737-5   ISBN 978-1-4614-4738-2 (eBook)
DOI 10.1007/978-1-4614-4738-2
Springer New York Heidelberg Dordrecht London

Library of Congress Control Number: 2012941620

Printed on acid-free paper

Springer is part of Springer Science+Business Media (www.springer.com)

Preface

In this book, I highlight the developments in Kalman filtering subject to general linear constraints. Essentially, the material to be presented is almost entirely based on the results and examples originally developed in Pizzinga et al. (2008a), Cerqueira et al. (2009), Pizzinga (2009, 2010), Souza et al. (2011), Pizzinga et al. (2011), and Pizzinga (2012). There are fundamentally three kinds of topics: (a) new proofs for already established results within the restricted Kalman filtering literature; (b) additional results that are should shed light on theoretical and methodological frameworks for linear state space modeling under linear restrictions; and (c) applications in investment analysis and in macroeconomics, where the proposed methods are illustrated and evaluated. At the end, I briefly discuss some extensions in the subject, which, again, step into theory, methods, and applications.

It is important to mention that my doctoral thesis, of which this book is a major revision, would not have been completed without the financial support from CNPq and FAPERJ. I would like to thank my friends and colleagues who have been important in my professional and personal life. I'd rather not list each one of them here, because they know who they are – and I refuse to run the risk of failing to mention someone.

I am especially indebted to certain professors who have greatly influenced and furthered my education and professional success. Some of them have become good friends over the years. In alphabetical order, I am grateful to Adherbal Filho, Antonio Dias, Ali Messaoudi, Carlos Kubrusly, Claudio Landim, Cristiano Fernandes, Edson Relvas, Eduardo Campos, Fernando Albuquerque, Frederico Cavalcanti, Hedibert Lopes, Inez Costa Chaves, Kaizo Beltrao, Leonardo Rolla, Luiz Carlos da Rocha, Marcelo Medeiros, Nei Carlos Rocha, Sergio Volchan, Waldir Lobao, and Zelia Bianchini.

Very special thanks go to my friends and colleagues from the Institute of Mathematics and Statistics of Fluminense Federal University. These folks gave me such a warm welcome... Honestly, when I am at work, I feel as if I am home.

Finally, I am grateful to my parents, Rose Nanie Heringer da Silva and Rodolfo Domenico Pizzinga, for their support. Furthermore, I must dedicate this book to my mother, to whom I am eternally indebted for her remarkable help and understanding throughout my life – and also for her invaluable support on proofreading my scientific texts, including this book.

Rio de Janeiro, Brazil   Adrian Pizzinga

Contents

1 Introduction 1
1.1 Motivation 1
1.2 A Glimpse at the Literature 2
    1.2.1 Statistics Papers 2
    1.2.2 Engineering Papers 3
1.3 The Book's Contents 4
1.4 Organization 5

2 Linear State Space Models and Kalman Filtering 7
2.1 The Model 7
2.2 Kalman Equations 7
2.3 Introducing Linear Restrictions 8

3 Restricted Kalman Filtering: Theoretical Issues 11
3.1 Augmented Restricted Kalman Filtering: Alternative Proofs 11
    3.1.1 Geometrical Proof 11
    3.1.2 Computational Proof 13
    3.1.3 Conditional Expectation Proof 15
3.2 Statistical Efficiency 16
3.3 Restricted Kalman Filtering Versus Restricted Recursive
     Least Squares 18
3.4 Initialization 21
    3.4.1 Motivation 21
    3.4.2 Reviewing the Initial Exact Kalman Smoother 21
    3.4.3 Combining Exact Initialization with Linear Restrictions 22

4 Restricted Kalman Filtering: Methodological Issues 27
4.1 Random-Walk State Vectors Under Time-Invariant Restrictions 27
4.2 Reduced Restricted Kalman Filtering 28
    4.2.1 Motivation 28
    4.2.2 The Method 29
    4.2.3 Reducing Versus Augmenting 30
4.3 Predictions from a Restricted State Space Model 32

5   Applications   35
   5.1   Case I: Semistrong Dynamic Style Analysis   36
       5.1.1   Motivation   36
       5.1.2   Competing Models   37
       5.1.3   Model Selection   39
       5.1.4   Empirical Results   40
   5.2   Case II: Estimation of Dynamic Exchange-Rate Pass-Through   44
       5.2.1   Motivation   44
       5.2.2   Empirical Results   46
   5.3   Case III: GDP Benchmarking Estimation and Prediction   51
       5.3.1   Motivation   51
       5.3.2   Model Setup   51
       5.3.3   Empirical Results   52

6   Further Extensions   53

References   55

List of Figures

Fig. 5.1 US dollar/real volatility under the AR(1)-GARCH(1,1) model 41
Fig. 5.2 Smoothed $STAR$ exposures and Jensen's measure for
          HSBC FIF Cambial with respective 95% confidence intervals....... 43
Fig. 5.3 Smoothed $STAR$ exposures and Jensen's measure
          for Itau Matrix US Hedge FIF with respective 95%
          confidence intervals 43
Fig. 5.4 IPA smoothed betas... 46
Fig. 5.5 IPA long -un exchange-rate pass-through 47
Fig. 5.6 IPA consumption-smoothed beta 48
Fig. 5.7 IPA production-smoothed beta.. 49
Fig. 5.8 IPC smoothed betas 50
Fig. 5.9 IPC long-run pass-through 50

List of Tables

Table 5.1   Results from estimations with HSBC FIF Cambial 41

Table 5.2   Results from the estimations with Itau Matrix US
Hedge FIF 42

Table 5.3   IPA information criteria of unrestricted and restricted models 47

Table 5.4   IPA estimated parameters, with corresponding
p-values in parentheses 47

Table 5.5   IPA consumption-estimated parameters, with
corresponding p-values in parentheses 48

Table 5.6   IPA production-estimated parameters, with
corresponding p-values in parentheses 49

Table 5.7   IPC information criteria of unrestricted and restricted models 50

Table 5.8   Results of benchmarking prediction 52

Chapter 1
Introduction

1.1 Motivation

In some relevant practical situations in areas such as economics, finance, actuarial science, and engineering, some state space models (Harvey 1989; Brockwell and Davis 1991; and Durbin and Koopman 2001) would make more sense if they were estimated under some meaningful restrictions on the state vector.

Some examples:

- One may consider state space models to conduct time-varying econometric models where well-established economic restrictions on the coefficients should be at least attested.
- Statistical models generated from physical considerations sometimes make sense only if considered under symmetry constraints on their parameters, whenever they are fixed or stochastically varying (cf. Pizzinga et al. 2005).
- In the claims reserving problem, some dynamic models for *runoff data* (cf. de Jong and Zehnwirth 1983) may have columns or nonnegativity restrictions in the development/delay effect.
- Dynamic factor models for portfolio on-line recovery should be at least subject to accounting restrictions (e.g., the portfolio allocations must add up to one for every time period; cf. Pizzinga and Fernandes 2006).

And whenever one attempts to perform such constrained state estimation, some questions naturally arise. How should this constrained estimation be implemented? When should this estimation be done? Which statistical properties do these methods of estimation share? Which theoretical or computational complications could emerge? Can all possible types of restrictions be handled? Can the imposed restrictions be checked for their plausibility under a specific method? And what could be said about the initialization of the recursions from these restricted estimations?

This book focuses on methods concerning *restricted Kalman filtering* appropriate to problems that require *linear* restrictions in the setting of *linear* state space models.

A. Pizzinga, *Restricted Kalman Filtering: Theory, Methods, and Application*,
SpringerBriefs in Statistics 12, DOI 10.1007/978-1-4614-4738-2_1,
© Springer Science+Business Media New York 2012

And the foregoing questions are addressed by a thorough analysis of theoretical results and methods, as well as by illustrative applications.

In the sequel, every topic discussed is examined in detail. But, first, without assuring complete coverage, let me review some literature on the subject.

1.2   A Glimpse at the Literature

Basically, the literature on linear state space models under restrictions has taken two directions – one is more "statistical-like" and the other is more "engineering-like." Since cross-references between these two fields have been rare, there is some overlap between contributions coming from both "worlds."

1.2.1   Statistics Papers

From a statistics/econometric standpoint, Doran (1992) is a seminal paper on the subject in which restricted Kalman filtering by augmentation was proved, under intense use of matrix algebra, for update and smoothing equations. At the end of his paper, Doran also made an attempt to further extend his approach to cases of nonlinear restrictions, but at least first-order differentiable. Over the ensuing 5 years, Doran published two other papers. In Doran (1996), his previous approach was used in a problem of estimating Australian provincial populations according to the annual national population. And in Doran and Rambaldi (1997), the same approach was once more evoked to solve the problem of estimating time-varying econometric models (demand systems to be exact) also under time-varying and quite interpretable restrictions; in that same paper, the authors also discussed the relevant question concerning numerical optimization for the maximum-likelihood estimation of unknown parameters. These three important papers – mainly the first one – have been cited regularly in the literature, and Doran's approach has been revisited as well; see, for instance, the book by Durbin and Koopman (2001), Sect. 6.5.

Other works on the subject are the papers by Pandher (2002, 2007), who also cited Doran in his bibliographic review. In those papers, Pandher was fully concerned about forecasting multivariate time series under linear restrictions. His approach is different from the one proposed by Doran, although some augmentation strategy under a structural modeling framework can be noted again. In the first of his two papers, Pandher gave some results on the statistical efficiency of state prediction under restrictions and on the observability of the state vector under his method. Lastly, Pandher recognized a previous and relevant work when he cited Leybourne (1993), who had tackled univariate state space models under time-invariant restrictions over a random-walk state vector.

Five recent papers in the statistics literature are Pizzinga et al. (2008a), Pizzinga (2009, 2010), Koop et al. (2010), and Pizzinga (2012). In Pizzinga et al. (2008a),

Doran's augmenting approach was demonstrated under elementary Hilbert space geometry without any Kalman equation, an approach that led to great generality on the types of linear constraints and of state vector smoothing. Pizzinga (2009) attempted to contribute to the theme by (a) giving another proof of Doran's approach, based entirely on quite elementary matrix operations; (b) establishing the statistical efficiency of the constrained Kalman filtering and smoothing under a purely geometrical perspective; and (c) suggesting an alternative approach to dealing with time-invariant constraints over random-walk state vectors. In Pizzinga (2010), the theme focused on four topics: the constrained Kalman filtering versus the recursive restricted least squares estimator; a new proof of the constrained Kalman filtering under a conditional expectation framework; linear constraints under a reduced state space modeling; and state vector prediction under linear constraints. Koop et al. (2010), instead, focused on Bayesian methods, where the posterior probability that a linear restriction holds at a particular time instant, given all the information available from the data set, is tackled. The authors also provided ways of generalizing their approach to cases where the restrictions are nonlinear or involve more than a time instant. Finally, Pizzinga (2012) investigated how the use of an initial diffuse state vector affects the use of the Kalman smoother under linear restrictions. It was established that it is still possible to obtain restricted smoothed state vectors in the "diffuse" period under quite general conditions, and this extension of the restricted Kalman smoother proved to preserve the conditional statistical efficiency.

1.2.2   Engineering Papers

Now, let us take a closer look at engineering articles about constrained Kalman filtering. First, there are the papers by Massicotte et al. (1995), who proposed a method for imposing positivity constraints, and by Geeter et al. (1997), in which a smoothly constrained Kalman filter for nonlinear constraints was developed. Stepping further, Simon and Chia (2002) derived at least two different versions of constrained Kalman filtering by alternative perspectives, all of them remaining on the Lagrange multipliers approach. Their first version of Kalman filtering is, in fact, quite similar to that originally obtained by Doran (1992) for the updating equations; this is only one example of the aforementioned overlap found in the literature. Simon and Chia presented five theorems that reveal good properties of their developed constrained Kalman filtering; four of them are related to mean square error efficiency. The methodological/theoretical part of that paper was closed by a discussion on how nonlinear identities could be encompassed by their constrained Kalman filtering. As a continuation, Simon and Simon (2004) tried to incorporate inequality constraints in Kalman filtering. The authors accomplished the task using quadratic programming, and it is interesting that the problem actually remains as a special case of imposing equality constraints.

More recently, Julier and Laviola (2007) concentrated almost entirely on non-linear constraints. The authors developed a two-stage state estimation under fairly general equality constraints. In that same year, Ko and Bitmead (2007) developed an alternative and quite intriguing approach to dealing with time-invariant and "homogeneous" linear constraints on a state vector. Their resulting constrained Kalman filtering was duly compared, in terms of statistical efficiency, with the unconstrained Kalman equations and with the best of all proposals from Simon and Chia (2002); under some theoretical conditions, Ko and Bitmead have established the superiority of their approach, which was illustrated by a specific numerical example.

Finally, the two most recent works were both published in 2009. First, the paper by Teixeira et al. (2009), whose contributions are as follows: (a) the recognition that linear constraints can arise from a reduction in the rank of one of the system matrices; (b) the derivation of a constrained Kalman filtering from the viewpoint of a maximum-a-posteriori solution; (c) the connections between their constrained Kalman filtering, the approach by Ko and Bitmead (2007), the approach by Simon and Chia (2002), and the augmented model approach raised from Theorem 3.1; and (d) a treatment of nonlinear constraints with generalizations of a well-established method for dealing with a nonlinear state space model, namely, *unscented Kalman filtering* (Julier and Uhlmann 2004). Second, the survey offered by Simon (2009) on several ways of imposing constraints, whether linear or nonlinear, on the state vector estimation. Simon discussed many of the approaches revisited in this subsection, providing detailed descriptions and numerical examples of the methods considered in his survey.

1.3 The Book's Contents

The specific contributions of this book consist of gathering most of the theoretical developments offered in Pizzinga et al. (2008a) and Pizzinga (2009, 2010, 2012). Also, three applications, previously tackled by Cerqueira et al. (2009), Pizzinga (2010), Souza et al. (2011), and Pizzinga et al. (2011), are presented and discussed as well. To be more precise, the focus is on the investigation and development of the following topics:

1. A more general and elegant proof of restricted Kalman filtering (updating and smoothing equations) that uses *Hilbert space geometry*. This proof is compared with those from Doran (1992).
2. An alternative proof based on Kalman recursions of restricted Kalman filtering. Its importance lies in the idea of *rewriting the augmented model* in a useful and equivalent way, which would be the building block for other methods and results within restricted Kalman filtering.
3. Two proofs of the *statistical efficiency* from restricted Kalman filtering.

4. An alternative approach to imposing time-invariant restrictions to the estimation of random-walk state vectors.
5. A comparison between restricted Kalman filtering and the restricted *recursive least squares*, and the establishment of the equivalence between both techniques under a particular, albeit relevant, case.
6. Development and implementation of a new restricted Kalman filtering under a *reduced modeling approach*. Such a method, originally proposed by Doran and Rambaldi (1997), is directly compared with the usual restricted Kalman filtering by augmentation.
7. A *restricted Kalman predictor* applicable to general situations that encompasses the method by Pandher (2002) as a particular case.
8. An alternative, "parametric," very short, and quite general proof of restricted Kalman filtering under a *conditional expectation* framework, followed by comparisons with previous demonstrations.
9. The proof that the initial exact Kalman smoother (cf. Durbin and Koopman 2001, Chap. 5) still yields restricted smoothed state vectors within the "diffuse" period whenever applied to an appropriate augmented model.
10. A practical illustration in finance, in which a dynamic factor model under a linear and interpretable restriction is used to understand the style of Brazilian exchange rate funds.
11. An application in macroeconomics, in which dynamic models for exchange rate pass-through are proposed and estimated using Brazilian price indexes.
12. A practical illustration in macroeconomics, in which a univariate benchmarking model, recognized as a linear state space model under restrictions, is used to predict Brazil's Gross Domestic Product.

The foregoing topics could be classified into the following three groups:

• Topics 1–3, 5, 8, and 9 full under *theory*.
• Topics 4, 6, and 7 cover *methods*.
• Topics 10–12 offer three *applications*.

1.4   Organization

This book is organized as follows. Chapter 2 revisits the essentials of linear state space models and Kalman filtering. Chapter 3 is totally dedicated to the theoretical issues concerning the imposition of linear restrictions on the Kalman equations, in which alternative proofs for already established results are given and, in addition, some additional results are derived. Chapter 4 focuses on some methods that can be potentially useful in situations of linear state space modeling under linear restrictions on the state vector. Chapter 5 offers the aforementioned applications in finance and macroeconomics, which illustrate the performance of some methods discussed in Chap. 4. Finally, Chap. 6 closes the book by suggesting some additional research topics.

Chapter 2
Linear State Space Models and Kalman Filtering

2.1 The Model

A *linear wide-sense* state space model for an observable p-variate stochastic process Y_t, defined on an appropriate probability space $(\Omega, \mathcal{F}, \mathcal{P})$, is described by the following set of equations:

$$Y_t = Z_t \alpha_t + d_t + \varepsilon_t,$$
$$\alpha_{t+1} = T_t \alpha_t + c_t + R_t \eta_t. \tag{2.1}$$

The first equation is usually called the *measurement equation*, and the second is known as the *state equation*. The unobservable m-variate process α_t is termed the *state vector* and is such that $E(\alpha_1) = a_1$ and $Var(\alpha_1) = P_1$. The error terms ε_t and η_t are respectively p-variate and r-variate second-order processes that are uncorrelated in time and from each other, with $var(\varepsilon_t) = H_t$ and $var(\eta_t) = Q_t$. The remaining *system matrices* $Z_t, d_t, H_t, T_t, c_t, R_t$, and Q_t evolve deterministically.

2.2 Kalman Equations

In this book, I will adopt the following notation:

- $a_{t|j}$ is an (equivalence class of) random vector(s) with coordinates $a_{ti|j}$, $i = 1, \ldots, m$, representing the unique linear orthogonal projection (cf. Kubrusly 2001, Theorem 5.52), evaluated on each (equivalence class of) coordinate(s) α_{ti} of α_t, onto $S' \equiv span\{1, Y_{11}, \ldots, Y_{1p}, \ldots, Y_{j1}, \ldots, Y_{jp}\} \subseteq L_2 \equiv L_2(\Omega, \mathcal{F}, \mathcal{P})$; the subjacent topology is that induced by the usual inner product, which is given by

$$< X, Y > \equiv E(XY) = \int_\Omega X(\omega)Y(\omega)\mathcal{P}(d\omega), \forall X, Y \in L_2;$$

A. Pizzinga, *Restricted Kalman Filtering: Theory, Methods, and Application*,
SpringerBriefs in Statistics 12, DOI 10.1007/978-1-4614-4738-2_2,
© Springer Science+Business Media New York 2012

- $P_{t|j} \equiv E\left[(\alpha_t - a_{t|j})(\alpha_t - a_{t|j})'\right]$;
- $v_t \equiv Y_t - Z_t a_{t|t-1} - d_t$ (this is the *innovation vector*) and $F_t \equiv E(v_t v_t') = Z_t P_{t|t-1} Z_t' + H_t$.

Kalman filtering (prediction, updating, and smoothing) gives the preceding orthogonal projection evaluations and the corresponding mean square error matrices. The corresponding equations are given as follows:

- Prediction equations:

$$a_{t+1|t} = T_t a_{t|t} + c_t,$$
$$P_{t+1|t} = T_t P_{t|t} T_t' + R_t Q_t R_t'. \tag{2.2}$$

- Updating or filtering equations:

$$a_{t|t} = a_{t|t-1} + P_{t|t-1} Z_t' F_t^{-1} v_t,$$
$$P_{t|t} = P_{t|t-1} - P_{t|t-1} Z_t' F_t^{-1} Z_t P_{t|t-1}. \tag{2.3}$$

- Smoothing equations (for a given $n \geq t$):

$$a_{t|n} = a_{t|t-1} + P_{t|t-1} r_{t-1},$$
$$r_{t-1} = Z_t' F_t^{-1} v_t + (T_t - T_t P_{t|t-1} Z_t' F_t^{-1} Z_t)' r_t,$$
$$P_{t|n} = P_{t|t-1} - P_{t|t-1} N_{t-1} P_{t|t-1},$$
$$N_{t-1} = Z_t' F_t^{-1} Z_t + (T_t - T_t P_{t|t-1} Z_t' F_t^{-1} Z_t)' N_t (T_t - T_t P_{t|t-1} Z_t' F_t^{-1} Z_t),$$
$$r_n = 0 \text{ and } N_n = 0. \tag{2.4}$$

Details concerning the derivations of these formulae are found in Harvey (1989), de Jong (1989), Brockwell and Davis (1991), Harvey (1993), Hamilton (1994), Tanizaki (1996), Durbin and Koopman (2001), Brockwell and Davis (2003), and Shumway and Stoffer (2006).

2.3   Introducing Linear Restrictions

Henceforth it is assumed that the process α_t in (2.1) satisfies linear restrictions as follows:

Assumption 2.1. *The random vectors α_t satisfy the following (possibly time-varying) linear restrictions:*

$$A_t \alpha_t = q_t, \tag{2.5}$$

where, for each t, A_t is a $k \times m$ matrix and q_t is a $k \times 1$ (possibly random) vector.

Observe that the restrictions enunciated in Eq. (2.5) are rather general. In fact, it encapsulates *affine* restrictions of the kind $A_t \alpha_t + b_t = q_t$ by defining $q'_t = q_t - b_t$ and allows the number of restrictions k to be time-varying. In practical situations, justification of such constraints in (2.5) arises naturally from the characteristics of the problem being modeled; see, for instance, the restrictions imposed on a demand system problem in Doran and Rambaldi (1997).

In the remainder of the book, Assumption 2.1 will be considered in almost every topic to be discussed and, in due course, may be added with some further structure.

Chapter 3
Restricted Kalman Filtering: Theoretical Issues

This entire chapter will be devoted to a discussion of several topics concerning the theory of imposing linear restrictions enunciated under a quite general form in (2.5) from Assumption 2.1. In Sect. 3.1, I will present and compare three different derivations of the restricted Kalman updating and smoothing equations under an augmented modeling approach. In Sect. 3.2, the statistical efficiency due to the imposition of restrictions is proved, and this shall be done using a geometrical framework. Moving forward, I try in Sect. 3.3 to establish the equivalence between restricted Kalman filtering and something that could be termed a *recursive restricted least squares* estimator. Finally, in Sect. 3.4, I investigate how initial diffuse state vectors affect the use of the Kalman smoother under linear restrictions.

3.1 Augmented Restricted Kalman Filtering: Alternative Proofs

3.1.1 Geometrical Proof

When estimating state space models under linear restrictions as given in Eq. (2.5), the natural task is to impose these very restrictions on the state estimators given by the Kalman equations to obtain a more meaningful result. The following theorem guarantees that such a task is possible for the updating and smoothing equations whenever one adopts an augmented measurement equation:

Theorem 3.1. *If the measurement vectors Y_t are replaced by $Y_t^* = (Y_t', q_t')'$, the matrices Z_t are replaced by $Z_t^* = [Z_t' \ A_t']'$, the vectors d_t are replaced by $d_t^* = (d_t', 0')'$, and the measurement equation error vectors ε_t are replaced by $\varepsilon_t^* = (\varepsilon_t', 0')'$, then the Kalman updating and smoothing equations applied to the new linear state space models satisfy the same linear restrictions given in (2.5), that is,*

A. Pizzinga, *Restricted Kalman Filtering: Theory, Methods, and Application*, SpringerBriefs in Statistics 12, DOI 10.1007/978-1-4614-4738-2_3, © Springer Science+Business Media New York 2012

$$A_t a_{t|t} = q_t, \tag{3.1}$$

$$A_t a_{t|n} = q_t. \tag{3.2}$$

First proof of Theorem 3.1. Denote the subspace generated by the augmented measurements up to time j, where $j \in \{t, t+1, \dots, n\}$, by

$$S'' = \text{span}\{1, Y_{11}, \dots, Y_{1p}, q_{11}, \dots, q_{1k}, \dots, Y_{j1}, \dots, Y_{jp}, q_{j1}, \dots, q_{jk}\},$$

the unique linear orthogonal projection onto S'' by $\pi_{S''}$, and the ith row from A_t by $A_{ti} = [c_{ti1} \dots c_{tim}]$. Then, making use of the linearity of $\pi_{S''}$ and of the linear restrictions established in Assumption 2.1, it follows that

$$A_{ti} a_{t|j} = c_{ti1} a_{t1|j} + \cdots + c_{tim} a_{tm|j} = c_{ti1} \pi_{S''}(\alpha_{t1}) + \cdots + c_{tim} \pi_{S''}(\alpha_{tm})$$

$$= \pi_{S''}(c_{ti1}\alpha_{t1} + \cdots + c_{tim}\alpha_{tm}) = \pi_{S''}(A_{ti}\alpha_t) = \pi_{S''}(q_{ti})$$

$$= q_{ti},$$

where the last equality comes from the fact that q_{ti} belongs to $\mathcal{R}(\pi_{S''}) = S''$. Since i is arbitrary, the theorem is proved. $\qquad\qquad\qquad\qquad\qquad\qquad\qquad\qquad\square$

Theorem 3.1 was originally due to Doran (1992, pp. 570 and 571), but the proof presented above, which was given in Pizzinga et al. (2008a), also reveals some gains:

1. It does not presume that F_t is invertible for all t.
2. It unifies in a single argument both updating and (*any type of*) smoothing equations.
3. It does not make any explicit use of Kalman updating or smoothing equations.
4. It is a shorter and, consequently, more elegant proof.

Item 1 above states a methodological contribution of this proof, namely, the guarantee that the augmented measurement procedure is able to deal with *any* type of linear restriction. Many examples of restrictions that would decrease the rank of F_t are of a deterministic nature, whether they originate from economic theories or not (to be even more specific: consider for instance the *portfolio* accounting restriction in time-varying extensions of the asset class factor models due to Sharpe 1992). A second contribution, related to item 2, is that *any* set of state smoothing (e.g., the traditional *fixed-interval*, *fixed-point*, and *fixed-lag* estimators; cf. Anderson and Moore 1979) must yield restricted estimated state vectors.

The following consequence of Theorem 3.1 has already proved to be useful, once it had been conveniently used by Doran (1996) in a state space estimation of population totals.

Corollary 3.2. (*"Perfect measurements"*) *If some univariate equations of the measurement vector Y_t have errors with zero variance, then*

$$Z_{t2} a_{t|t} = Y_{t2} \quad and \quad Z_{t2} a_{t|n} = Y_{t2}, \tag{3.3}$$

where Z_{t2} is the block from Z_t that corresponds to the block Y_{t2} from Y_t whose coordinates have null variance errors.

Proof. It is enough to see that Y_t can be written as $[Y'_{t1}\ Y'_{t2}]'$ and that Z_t, in turn, can be written as $[Z'_{t1}\ Z'_{t2}]'$. Establishing that $A_t = Z_{t2}$ and $q_t = Y_{t2}$, Theorem 3.1 guarantees the desired result. $\qquad\square$

3.1.2 Computational Proof

In this subsection, I will consider the following structure in addition to the restrictions in (2.5):

Assumption 3.3. *The linear restrictions in (2.5) are such that the coordinates of q_t are linearly independent in L_2 and from $1, Y_{11}, \ldots, Y_{1p}, \ldots, Y_{t1}, \ldots, Y_{tp}$. Also, suppose that $F_t > 0$ for all t.*

For the Kalman updating and smoothing equations, it is in fact an attainable task, as stated by Theorem 3.1, to carry out Kalman filtering estimations under the preceding linear restrictions. Here, this is now proved by explicitly using updating and smoothing equations, though under strategies somewhat different from those tackled by Doran (1992).

Second proof of Theorem 3.1. Uncouple the augmented model by recognizing that, for all t, q_t is a "new" measurement vector that is observed "after" Y_t and "before" Y_{t+1}. This recognition leads to a new linear state space representation entirely equivalent to the augmented model. The measurement equation for this representation is defined by

$$Y_{t,j} = Z_{t,j}\alpha_{t,j} + d_{t,j} + \varepsilon_{t,j}, \ \varepsilon_{t,j} \sim \text{WN}(0, H_{t,j}). \qquad (3.4)$$

When $j = 1$, nothing is changed from the measurement equation from (2.1) of Sect. 2.1. But for $j = 2$ we must have

$$Y_{t,2} = q_t, \ Z_{t,2} = A_t, \ d_{t,2} = 0 \text{ and } H_{t,2} = 0. \qquad (3.5)$$

Regarding the state equation, notice that, for all t, $\alpha_{t,2} = \alpha_{t,1}$ and $\alpha_{t+1,1} = T_t\alpha_{t,2} + c_t + R_t\eta_t$, $\eta_t \sim (0, Q_t)$. Within this equivalent framework, it becomes possible to treat the imposition of the linear restriction in time t as a new update of the state vector. Consider the state updating equation given in (2.3), already applied to the preceding equivalent model for t fixed and $j = 2$:

$$a_{t,2|t,2} = a_{t|t-1,2} + P_{t|t-1,2}Z'_{t,2}F_{t,2}^{-1}(Y_{t,2} - Z_{t,2}a_{t|t-1,2})$$

$$= a_{t|t-1,2} + P_{t|t-1,2}Z'_{t,2}(Z_{t,2}P_{t|t-1,2}Z'_{t,2} + H_{t,2})^{-1}(Y_{t,2} - Z_{t,2}a_{t|t-1,2})$$

$$= a_{t|t-1,2} + P_{t|t-1,2} A_t' (A_t P_{t|t-1,2} A_t')^{-1} (q_t - A_t a_{t|t-1,2}),$$

where the second equality comes from the very expression of F_t (cf. the established notation in Sect. 2.2) and the third comes from (3.5). Now, since $(A_t P_{t|t-1,2} A_t')^{-1}$ is a genuine inverse (cf. Assumption 3.3) and $a_{t,2|t,2} = a_{t|t}$, this last updated state vector being the one associated with the augmented model, premultiply both sides of the last identity by A_t to obtain (3.1).

Now rephrase the state smoothing equations in (2.4) for the augmented model as follows:

$$a_{t|n} = a_{t|t-1} + P_{t|t-1} r_{t-1},$$

$$r_{t-1} = Z_t^{*'} F_t^{-1} v_t + \left(T_t - T_t P_{t|t-1} Z_t^{*'} F_t^{-1} Z_t^* \right)' r_t, \quad \text{where} \quad Z_t^* = \begin{bmatrix} Z_t \\ A_t \end{bmatrix}.$$

Of course, other quantities would also have deserved asterisks, but they are suppressed for ease of notation. Placing the expression of r_t in $a_{t|n}$, it follows that

$$a_{t|n} = a_{t|t-1} + P_{t|t-1} (Z_t^{*'} F_t^{-1} v_t + (T_t - T_t P_{t|t-1} Z_t^{*'} F_t^{-1} Z_t^*)' r_t)$$

$$= a_{t|t-1} + P_{t|t-1} Z_t^{*'} F_t^{-1} v_t + P_{t|t-1} (T_t - T_t P_{t|t-1} Z_t^{*'} F_t^{-1} Z_t^*)' r_t$$

$$= a_{t|t} + (P_{t|t-1} T_t' - P_{t|t-1} Z_t^{*'} F_t^{-1} Z_t^* P_{t|t-1} T_t') r_t,$$

where the last equality follows from the Kalman updating equation in (2.3). Premultiplying both sides by A_t, it follows that

$$A_t a_{t|n} = A_t a_{t|t} + (A_t P_{t|t-1} T_t' - A_t P_{t|t-1} Z_t^{*'} F_t^{-1} Z_t^* P_{t|t-1} T_t') r_t.$$

According to Doran (1992), Eq. (22) (from Assumption 3.3, F_t from the augmented model is invertible), $A_t P_{t|t-1} Z_t^{*'} F_t^{-1} = \begin{bmatrix} 0 & I \\ k \times p & k \times k \end{bmatrix}$. Use this together with (3.1), which was already proved, to obtain

$$A_t a_{t|n} = q_t + \left(A_t P_{t|t-1} T_t' - [0 \ I] \begin{bmatrix} Z_t \\ A_t \end{bmatrix} P_{t|t-1} T_t' \right) r_t$$

$$= q_t + \left(A_t P_{t|t-1} T_t' - A_t P_{t|t-1} T_t' \right) r_t = q_t,$$

which gives identity (3.2)   □

There is no methodological novelty here. In turn, the contribution offered comes from this second proof, which was offered in Pizzinga (2009) and deserves some qualification. Although it does not encompass significant generalizations like those verified in the first proof and is considerably longer, this second proof makes use of simple matrix operations, which illustrate potentially useful strategies that could be evoked in future research. Indeed, the same decomposition, which was used in that part of the proof that was related to the updating equations, has been equally important for the well-known treatment of multivariate state space models under a univariate framework (cf. Durbin and Koopman 2001, Sect. 6.4). In addition, linear state space models for benchmarking proposed by Durbin and Queenneville (1997) and later revisited by Durbin and Koopman (2001), Sect. 3.8, are based on such a rearrangement as well. On the other hand, the part related to the smoothing equation is entirely based on de Jong (1989)'s smoothing recursions, which are mathematically transparent and computationally efficient.

3.1.3 Conditional Expectation Proof

The main goal of this subsection is to give a third and final proof for augmented restricted Kalman filtering. For this, I must add another structure (quite "traditional," we would say) to the linear state space model in (2.1).

Assumption 3.4. ε_t and η_t are independent (in time, between each other, and of α_1) Gaussian stochastic processes. Also, α_1 is a Gaussian random vector.

In addition to considering this new "parametric" framework, denote by $\mathcal{F}_j$ the σ-field generated by the measurement vectors up to time j; that is, $\mathcal{F}_j \equiv \sigma\,(Y_1, \ldots, Y_j)$. Also, set $\hat{a}_{t|j} \equiv E\left(\alpha_t|\mathcal{F}_j\right)$ and $\hat{P}_{t|j} \equiv V\left(\alpha_t|\mathcal{F}_j\right)$. Under Assumption 3.4, the Kalman recursions are versions of these conditional moments when $j = t - 1$, $j = t$, and $j = n$; see Anderson and Moore (1979), Harvey (1989, 1993), Tanizaki (1996), and Durbin and Koopman (2001). Consequently, Pizzinga (2010) made use of some standard properties of the conditional expectation to obtain a very quick proof of Theorem 3.1.

Third proof of Theorem 3.1. Let t be an arbitrary time instant. Define $\mathcal{F}_j^* \equiv \sigma\left(Y_1, q_1, \ldots, Y_j, q_j\right)$. Fixing j in $\{t, t+1, \ldots, n\}$, it follows with probability 1 that

$$A_t \hat{a}_{t|j} = A_t E\left(\alpha_t|\mathcal{F}_j^*\right) = E\left(A_t \alpha_t|\mathcal{F}_j^*\right) = E\left(q_t|\mathcal{F}_j^*\right) = q_t, \qquad (3.6)$$

where the third equality is due to the restrictions in (2.5) and the fourth equality naturally comes from the very $\mathcal{F}_j^*$-measurability of q_t. Finally set $j = t$ and $j = n$. □

The most evident comparison between this third proof and the previous proofs is concerned with length and elegance. In addition, it maintains the same generality in terms of linear restrictions and state smoothing, which was guaranteed already by the first proof given in Sect. 3.1.1. There are two reasons for this smaller length: (a) $\sigma\left(Y_1, \ldots, Y_j\right) = \sigma\left(Y_{11}, \ldots, Y_{1p}, \ldots, Y_{j1}, \ldots, Y_{jp}\right)$ and (b) there is no lack of algebraic structure by taking conditional expectations of vectors of one particular dimension conditioned on σ-fields generated by vectors of entirely different dimensions.

Now, regarding the potential usefulness of this third proof:

1. The additional normality and independence assumptions, although slightly limiting the scope of Theorem 3.1, can be considered an asset because these are straightforwardly generalizable to other types of state space models – the non-Gaussian or nonlinear state space models. The only drawback is that most statistical techniques designed to handle more general state space models require the existence of an expression for the conditional laws $p(y_t|\alpha_t)$, which are obscured by the "singularity" incurred in the augmenting procedure.
2. Finally, the third proof plainly reveals that the *Bayesian approach* for state space modeling (cf. West and Harrison 1997; Durbin and Koopman 2001; and Shumway and Stoffer 2006) can also deal with linear restrictions by adopting augmented measurement equations as well. Yet, in one such case we will be aware of some unavoidable singularities.

3.2 Statistical Efficiency

In this section, the statistical efficiency – in terms of mean square estimation error – of the restricted Kalman filtering discussed so far is demonstrated. For this, I will make use of a geometrical perspective, something that might be general enough, while still grasping at intuition and simplicity. For what follows, it is important to bear in mind that Kalman recursions, in addition to being recursive computational formulae from an operational standpoint, give linear orthogonal projections evaluations on some specific subspaces spanned by the model measurements.

Let me begin by recalling the following well-established and useful fact:

Lemma 3.5. *Take a Hilbert space $\mathcal{H}$, two subspaces $\mathcal{M}, \mathcal{N}$ of $\mathcal{H}$, and the linear orthogonal projections $\pi_\mathcal{M}$ and $\pi_\mathcal{N}$. If $\mathcal{M} \subseteq \mathcal{N}$, then, for each $x \in \mathcal{H}$, $\pi_\mathcal{M}(\pi_\mathcal{N}(x)) = \pi_\mathcal{M}(x)$.*

Proof. $\mathcal{N}$ is, by itself, a Hilbert space (because it is closed) and $\mathcal{M}$ is a closed subspace of $\mathcal{N}$. Then, using the orthogonal projection theorem (cf. Theorem 5.20 of Kubrusly 2001), we obtain $\mathcal{N} = \mathcal{M} + (\mathcal{M}^\perp \cap \mathcal{N})$. Thus, from Proposition 5.58 of Kubrusly (2001) it follows that $\pi_\mathcal{N} = \pi_\mathcal{M} + \pi_{\mathcal{M}^\perp \cap \mathcal{N}}$; tautologically, for each $x \in \mathcal{H}$,

$$\pi_\mathcal{N}(x) = \pi_\mathcal{M}(x) + \pi_{\mathcal{M}^\perp \cap \mathcal{N}}(x). \tag{3.7}$$

Now, apply $\pi_\mathcal{M}$ to both sides of (3.7).   □

Some additional notation must also be set:

- $a_{t|j}$, $P_{t|j}$, and S' are defined as previously and relate to the *standard* state space model.
- $a^*_{t|j}$, $P^*_{t|j}$, and S'' are obtained with the *augmented* state space model associated with Theorem 3.1.

Now, everything needed for formally guaranteeing statistical efficiency has been gathered. Two demonstrations were given in Pizzinga (2009), and both are revisited here. These proofs rely on a strong geometrical appeal and have an inductive style, in the sense that, firstly, individual coordinates of the state vector are tackled and then, in a second moment, the strategy is generalized for arbitrary linear combinations of these coordinates. But they do differ in some aspects. The first proof concentrates on the optimality of the linear orthogonal projection that comes directly from first principles, while the second proof is rather "constructive," uses Lemma 3.5, and focuses on a standard decomposition.

Theorem 3.6. $P^*_{t|j} \leq P_{t|j}$ *in the usual ordering of symmetric matrices.*

First proof of Theorem 3.6. Let $i = 1, \ldots, m$. Since the set containing the original model measurements until time j is contained in the corresponding set from the augmented model, it follows that

$$S' \subseteq S'' \equiv \text{span}\{1, Y_{11}, \ldots, Y_{1p}, q_{11}, \ldots, q_{1k}, \ldots, Y_{j1}, \ldots, Y_{jp}, q_{j1}, \ldots, q_{jk}\}.$$

Therefore, from Theorem 5.53 of Kubrusly (2001),

$$E\left[\left(\alpha_{ti} - a^*_{ti|j}\right)^2\right] = \inf_{Y \in S''} E\left[(\alpha_{ti} - Y)^2\right] \leq E\left[\left(\alpha_{ti} - a_{ti|j}\right)^2\right].$$

Generalizing: take $x = (x_1, \ldots, x_m)' \in R^m$. Using linearity, the linear orthogonal projections onto S' and onto S'', both evaluated in $x'\alpha_t = x_1\alpha_{t1} + \cdots + x_m\alpha_{tm}$, are given by

$$x'a_{t|j} = x_1 a_{t1|j} + \cdots + x_m a_{tm|j}$$

and

$$x'a^*_{t|j} = x_1 a^*_{t1|j} + \cdots + x_m a^*_{tm|j}.$$

Observing that $x'a_{t|j} \in S''$ (because $a_{ti|j} \in S' \subseteq S'' \forall i = 1, \ldots, m$ and S'' is a linear manifold), it follows that

$$x' P^*_{t|j} x = x' E\left[(\alpha_t - a^*_{t|j})(\alpha_t - a^*_{t|j})'\right] x = E\left[x'(\alpha_t - a^*_{t|j})(\alpha_t - a^*_{t|j})'x\right]$$

$$= E\left[(x'\alpha_t - x'a^*_{t|j})(x'\alpha_t - x'a^*_{t|j})'\right] = E\left[\left(x'\alpha_t - x'a^*_{t|j}\right)^2\right]$$

$$= \inf_{Y \in S''} E\left[(x'\alpha_t - Y)^2\right] \leq E\left[(x'\alpha_t - x'a_{t|j})^2\right]$$

$$= E\left[(x'\alpha_t - x'a_{t|j})(x'\alpha_t - x'a_{t|j})'\right] = E\left[x'(\alpha_t - a_{t|j})(\alpha_t - a_{t|j})'x\right]$$

$$= x' E\left[(\alpha_t - a_{t|j})(\alpha_t - a_{t|j})'\right] x = x' P_{t|j} x.$$

Since x is arbitrary, the conclusion is that $P^*_{t|j}$ is, in fact, "less than or equal to" $P_{t|j}$.   □

Second proof of Theorem 3.6. Consider again an arbitrary $i = 1,\ldots,m$. Recall once more that S' and S'', already defined, are subspaces (closed linear manifolds) of L_2 and that $S' \subseteq S''$. Theorem 5.20 of Kubrusly (2001) asserts the existence of $\xi \in S''^{\perp}$ such that

$$\alpha_{ti} = a^*_{ti|j} + \xi. \tag{3.8}$$

Also, Theorem 5.20 of Kubrusly (2001) and Lemma 3.5 (make $\mathcal{H} = L_2$, $\mathcal{M} = S'$, $\mathcal{N} = S''$, and $x = \alpha_{ti}$) assure the existence of $v \in S'' \cap S'^{\perp}$ such that

$$a^*_{ti|j} = a_{ti|j} + v. \tag{3.9}$$

From the decomposition (3.8) we obtain

$$E\left[\left(\alpha_{ti} - a^*_{ti|j}\right)^2\right] = E\left(\xi^2\right). \tag{3.10}$$

Now, add both decompositions (3.8) and (3.9) to obtain $\alpha_{ti} - a_{ti|j} = \xi + v$, and evoke the Pythagorean theorem – which is licit since ξ and v are orthogonal – to obtain

$$E\left[\left(\alpha_{ti} - a_{ti|j}\right)^2\right] = E\left(\xi^2\right) + E\left(v^2\right). \tag{3.11}$$

Identities (3.10) and (3,11) assert the claimed efficiency for each coordinate estimation of the state vector. The case of an arbitrary linear combinations $x'\alpha_t$ is dealt with in a similar fashion.   □

Looking at cases in which $j \geq t$, the last theorem shows that Kalman updating and smoothing equations, when used with the augmented model, besides respecting the linear restrictions from Eq. (3.1), give more accurate estimators.

3.3 Restricted Kalman Filtering Versus Restricted Recursive Least Squares

Consider the following univariate special case of model (2.1) where the state vector is time invariant and $Z_t \equiv x'_t$ is a row vector of exogenous explanatory variables:

$$Y_t = x'_t\beta_t + \varepsilon_t, \quad \varepsilon_t \sim (0,\sigma^2),$$
$$\beta_{t+1} = \beta_t. \tag{3.12}$$

This model can and should be viewed as a linear regression model written in a linear state space representation. It is known (Harvey 1981, 1993) that application of the Kalman state updating equation to (3.12) numerically coincides with the method of recursive least squares. Back in the days when matrix inversion was a computational burden, this equivalence proved useful since it turned out to be possible to estimate a regression model with no need to invert a "big" $X'X$ matrix. In addition, this made attainable the updating of ordinary least squares (OLS) estimates whenever new observations were added to the data set. Nowadays this equivalence still deserves its merits in regression analysis and econometrics. Firstly, depending on the ill-conditioning of the regressors, it still may be difficult to invert "big" $X'X$ matrices, a problem that justifies recursive estimation. Secondly, this equivalence is in full connection with the traditional coefficient stability test of Brown et al. (1975).

The purpose of this section is to revisit the generalization of the foregoing parallel addressed by Pizzinga (2010) in the context of linear restrictions. Thus, it is assumed that the coefficient vector of a regression model is supposed to obey certain linear restrictions that are enunciated as

$$A\beta = q, \tag{3.13}$$

where A is a known $k \times m$ matrix, $k \leq m$, and $q = (q_1, \ldots, q_k)'$ is a known $k \times 1$ vector. Since the main objective is to bridge the restricted recursive estimation to the restricted Kalman filtering, structures (3.12) and (3.13) are now taken together to generate the following augmented measurement equation:

$$\begin{pmatrix} Y_t \\ q \end{pmatrix} = \begin{pmatrix} x_t' \\ A \end{pmatrix} \beta_t + \begin{pmatrix} \varepsilon_t \\ 0 \end{pmatrix}, \quad \begin{pmatrix} \varepsilon_t \\ 0 \end{pmatrix} \sim \left(\begin{pmatrix} 0 \\ 0 \end{pmatrix} \begin{pmatrix} \sigma^2 & 0 \\ 0 & 0 \end{pmatrix} \right). \tag{3.14}$$

From Theorem 3.1, the application of the Kalman updating equation to the model in (3.14) produces updated state vectors that satisfy $Ab_{t|t} = q$. But, in fact, there is more: the terms of the sequence $(b_{t|t})$ are the output from *online* successive applications of restricted least squares. To establish this link, the restricted least squares (RLS) estimator and its covariance matrix for a linear regression model $Y = X\beta + \varepsilon, \varepsilon \sim (0, \sigma^2 I)$, where β obeys (3.13), is recalled below:

$$\hat{\beta}_{RLS} = \hat{\beta}_{OLS} + (X'X)^{-1}A'[A(X'X)^{-1}A']^{-1}(q - A\hat{\beta}_{OLS}), \tag{3.15}$$

$$\text{Var}\left(\hat{\beta}_{RLS}\right) = \sigma^2 (X'X)^{-1} - (X'X)^{-1}A'[A(X'X)^{-1}A']^{-1}A(X'X)^{-1}.$$

The derivation of the expression in (3.15) is presented in virtually all books on econometrics. See, for instance, Johnston and Dinardo (1997) or Greene (2003).

This section's result:

Theorem 3.7. *Under the state space model in (3.14), the Kalman state updating equation is identical to a recursive application of (3.15).*

Proof. The model in (3.14) can be decomposed in such a way that recognizes q as a "new" measurement vector obtained/observed just "after" Y_t and right "before"

Y_{t+1}. Thus, the measurement equation is recast as

$$Y_{t,j} = Z_{t,j} \beta_{t,j} + \varepsilon_{t,j}, \quad \varepsilon_{t,j} \sim (0, H_{t,j}). \tag{3.16}$$

Notice that $Y_{t,1} = Y_t$, $Z_{t,1} = x_t'$ and $H_{t,1} = \sigma^2$; on the other hand, $Y_{t,2} = q$, $Z_{t,2} = A$, and $H_{t,2} = 0$. The new state equation is written in the same way as before.

It now becomes possible to regard imposing the linear restrictions as a new update of the state vector. In fact, for an arbitrary t, denote the output of the Kalman updating using all the measurements from Eq. (3.16) up to $Y_{t,1}$ by $\hat{\beta}_{t,1|t,1}$, which, as was already discussed, equals the output of the recursive least squares – and consequently the OLS estimator – applied to the "observations" $\{Y_1, q_1, \ldots, q_k, \ldots, Y_{t-1}, q_1, \ldots, q_k, Y_t\}$. The state equation implies that $\hat{\beta}_{t,2|t,1} = \hat{\beta}_{t,1|t,1}$. Then, as $Y_{t,2} = q$ arrives, and because $P_{t,2|t,1} = P_{t,1|t,1} = \sigma^2 \left(X_t' X_t\right)^{-1}$, $Z_{t,2} = A$ and $\upsilon_{t,2} = q - A\hat{\beta}_{t,2|t,1}$, and $F_{t,2} = A\sigma^2 \left(X_t' X_t\right)^{-1} A'$, the Kalman state updating equation in (2.3) becomes

$$\hat{\beta}_{t,2|t,2} = \hat{\beta}_{t,1|t,1} + \left(X_t' X_t\right)^{-1} A' \left(A \left(X_t' X_t\right)^{-1} A'\right)^{-1} \left(q - A\hat{\beta}_{t,1|t,1}\right). \tag{3.17}$$

But, as was just mentioned, $\hat{\beta}_{t,1|t,1} = \hat{\beta}_{MQO}$. Therefore, the conclusion is that the Kalman updating in (3.17) is indeed an application of the RLS estimator of (3.15). The equivalence between covariance matrices can be established analogously. □

Some conceptual and practical consequences follow from this proof, which was previously offered in Pizzinga (2010). First of all, it now becomes clear that restricted Kalman filtering is indeed a generalization of the RLS estimator, a statement that, albeit intuitive, requires proper formalization. In addition, Theorem 3.7 also shows that a regression model with random-walk time-varying coefficients *under restrictions* (set $\beta_{t+1} = \beta_t + \eta_t$, $\eta_t \sim$ WN $(0, Q)$, as the state equation for model (3.14)) does encapsulate the regression model with static coefficients, still under the same restrictions. Then, whenever restricted Kalman filtering is applied to the time-varying version, both models can be compared as usual – to estimate the static model, just set $Q \equiv 0$. Finally, note that the recursive residuals obtained from the recursive application of (3.15) are automatically uncorrelated – indeed, Theorem 3.7 says they are innovations. This is a desirable property in paving the way toward the development of a generalization of the stability test by Brown et al. (1975).

3.4 Initialization

3.4.1 Motivation

Besides considering the linear restrictions in Eq. (2.5), in this section it is assumed that some coordinates of the initial state vector α_1 have infinity variances. This is the basic setup of the so-called *diffuse initialization* of Kalman recursions, a subject extensively studied in Ansley and Kohn (1985), de Jong (1988), Harvey (1989), de Jong (1991), Koopman (1997), Durbin and Koopman (2001), Koopman and Durbin (2003), and de Jong and Chu-Chun-Lin (2003). Under these at least partially unspecified initial conditions, the question arises as to whether the methods of imposing linear restrictions can be derived from the very beginning. Observe that, once some elements of P_1 explode, there will no longer be an L_2 theory available, nor could the traditional Kalman equations even be tackled, at least in the period when the effect of diffuseness – which lasts for an initial portion of the data – has not yet vanished. Thus, the strategies used in proofs by Doran (1992), Pizzinga et al. (2008a), and Pizzinga (2009) – the latter two already revisited in this book in Sects. 3.1.1 and 3.1.2, respectively – for augmented restricted Kalman filtering (cf. Theorem 3.1) unfortunately become useless now. The purpose of this section is to address this theoretical issue precisely, much in the same way as was done in Pizzinga (2012): exploring the conditions that allow one to extend the restricted estimation to diffuse initializations and by working out appropriately the modified versions of the Kalman equations, that is, the proof of the main result will be "computational" instead of "geometrical."

3.4.2 Reviewing the Initial Exact Kalman Smoother

Henceforth, the initial state vector is modeled as

$$\alpha_1 = a + B\delta + R_0\eta_0,$$

where (a) a is known; (b) δ and η_0 are zero-mean random vectors such that $\text{Var}(\delta) = \kappa I_q$, $\text{Var}(\eta_0) = Q_0$, and $\text{Cov}(\delta, \eta_0) = 0$; and (c) B and R_0 are $m \times q$ and $m \times (m-q)$ selection matrices, respectively, such that $B'\alpha_1 = \delta$. In general, δ consists of initial conditions for the nonstationary terms of the m-variate process α_t. Under this fix and assuming $F_t > 0$, the exact initial Kalman filter and smoother, obtained when $\kappa \longrightarrow +\infty$, are, in the notation of Durbin and Koopman (2001),

$$v_t^{(0)} = Y_t - Z_t a_t^{(0)} - d_t, \quad F_{*,t} = Z_t P_{*,t} Z_t' + H_t,$$

$$F_{\infty,t} = Z_t P_{\infty,t} Z_t', \qquad L_t^{(0)} = T_t - T_t P_{\infty,t} Z_t' F_{\infty,t}^{-1} Z_t,$$

$$L_t^{(1)} = -T_t P_{*,t} Z_t' F_{\infty,t}^{-1} Z_t + T_t P_{\infty,t} Z_t' F_{\infty,t}^{-1} F_{*,t} F_{\infty,t}^{-1} Z_t,$$

$$a_{t+1}^{(0)} = T_t a_t^{(0)} + c_t + T_t P_{\infty,t} Z_t' F_{\infty,t}^{-1} v_t^{(0)},$$

$$(3.18)$$

$$P_{*,t+1} = T_t P_{\infty,t} L_t^{(1)'} + T_t P_{*,t} L_t^{(0)'} + R_t Q_t R_t',$$

$$P_{\infty,t+1} = T_t P_{\infty,t} L_t^{(0)'}, \quad t = 1, \ldots, n,$$

$$r_{t-1}^{(0)} = L_t^{(0)'} r_t^{(0)}, \quad r_{t-1}^{(1)} = Z_t F_{\infty,t}^{-1} v_t^{(0)} + L_t^{(0)'} r_t^{(1)} + L_t^{(1)'} r_t^{(0)},$$

$$a_{t|n} = a_t^{(0)} + P_{*,t} r_{t-1}^{(0)} + P_{\infty,t} r_{t-1}^{(1)}, \quad r_n^{(0)} = 0, \quad r_n^{(1)} = 0, \quad t = n, \ldots, 1,$$

$$(3.19)$$

whenever $F_{\infty,t}$ just defined is nonsingular. Otherwise, changes must take place in recursions (3.18) and (3.19) (cf. Koopman and Durbin 2003). According to Koopman (1997), there exists a time instant d after which the preceding recursions collapse to the traditional Kalman smoother; therefore, $P_{\infty,t} = 0$ for $t > d$ necessarily. The set of time instants $\mathcal{D} \equiv \{1, \ldots, d\}$ is termed the *diffuse period*.

The presented recursions constitute the paradigm proposed in Koopman (1997), Durbin and Koopman (2001, Sect. 5.3), and Koopman and Durbin (2003) for the treatment of state smoothing diffuse initialization. For an alternative approach, based on the augmentation of the measurement equation, see de Jong and Chu-Chun-Lin (2003).

3.4.3 Combining Exact Initialization with Linear Restrictions

Before going to the main result of this Sect. 3.4, some preliminary steps must be addressed. The first is to list and to discuss the conditions under which it will be possible to combine diffuse initial conditions for the Kalman smoothing equations with the imposition of linear restrictions. Let them be enunciated and, without any loss of generality, consider them valid for all $t = 1, \ldots, n$.

Assumption 3.8. $\{q_{ti}, \ldots, q_{tk}\}$ *is a linearly independent set of observable random variables.*

Assumption 3.9. $\forall i = 1, \ldots, k : q_{ti} \notin \text{span}\{1, Y_{11}, \ldots, Y_{1p}, q_{11}, \ldots, q_{1k}, \ldots,$
$Y_{t-1,1}, \ldots, Y_{t-1,p}, q_{t-1,1}, \ldots, q_{t-1,k}, Y_{t,1}, \ldots, Y_{t,p}\}$.

Assumption 3.10. $\forall i = 1, \ldots, p : \text{row}_i(Z_t) \notin \text{span}\{\text{row}_1(A_t), \ldots, \text{row}_k(A_t)\} \subseteq \Re^m$.

Assumption 3.11. *Range* $(\alpha_t) = \Re^m$.

Assumption 3.12. $P_{\infty,t} \neq 0 \Rightarrow A_t P_{\infty,t} A_t' > 0$.

For a full discussion about the generality and plausibility of the preceding assumptions, the reader is referred to Pizzinga (2012). But, essentially, Assumptions 3.8–3.12 guarantee that the linear restrictions are nonredundant, observable, and handleable by augmented restricted Kalman filtering (cf. Theorem 3.1); in addition, they assure that Eqs. (3.18) and (3.19) will always be valid for state vector estimation in the diffuse period.

Before we proceed, three auxiliary results, whose proofs are given in Pizzinga (2012) and Doran (1992), are needed. The first is basic for embedding the problem of imposing restrictions on the theory of the exact initial Kalman filter.

Lemma 3.13. *Consider a state space model with the augmented measurement equation suggested in Theorem 3.1 and a finite initial covariance matrix P_1. Under Assumptions 3.8 and 3.9, $A_t P_{t|t-1} A_t' > 0$.*

The second result will guarantee that no other versions of the exact initial Kalman filter and smoother, unlike those given in (3.18) and (3.18) (which assume $F_{\infty,t} > 0$), will be required.

Lemma 3.14. *Consider a state space model with the augmented measurement equation suggested in Theorem 3.1 and a finite initial covariance matrix P_1. Under Assumptions 3.10 and 3.11, $A_t P_{t|t-1} A_t' > 0$ implies $\left[Z_{t,i}'\ A_t' \right]' P_{t|t-1} \left[Z_{t,i}'\ A_t' \right] > 0$ for each $i = 1, \ldots, p$, where $Z_{t,i} \equiv row_i (Z_t)$.*

The third result is actually a rephrasing of Eq. (22) from Doran (1992), which has already proved to be key in other situations concerning the theory of state space models under restrictions (see Pizzinga 2009).

Lemma 3.15. *Consider a state space model with the augmented measurement equation suggested in Theorem 3.1 and a finite initial covariance matrix P_1. If $F_t > 0$, then $A_t P_{t|t-1} \left[Z_t'\ A_t' \right] F_t^{-1} = \left[0_{k \times p}\ I_{k \times k} \right]$.*

Finally, here is the main result concerning initialization:

Theorem 3.16. *("The initial exact restricted Kalman smoother") Suppose the augmented state space model suggested in Theorem 3.1 satisfies Assumptions 3.8– 3.12. Then the initial exact Kalman smoother in (3.19) yields*

$$A_t a_{t|n} = q_t. \tag{3.20}$$

Proof. Consider first the case of $t \in \mathcal{D}$, where $\mathcal{D}$ corresponds to the augmented model. From Lemma 3.13, prior to letting $\kappa \longrightarrow \infty$ the covariance matrix F_t associated with the augmented model is positive definite, which assures that the exact initial Kalman filter and smoother can be considered in what follows. Define $\tilde{Z}_t = \left[Z_t'\ A_t' \right]'$. Other quantities would have also deserved tildes, but they are dropped to conserve notation. Define an auxiliary (augmented) state space model with system matrices given by $Z_t^\dagger = \tilde{Z}_t$, $d_t^\dagger = \left(d_t'\ 0' \right)'$, $H_t^\dagger = 0_{(p+k) \times (p+k)}$, $T_t^\dagger = T_t$, $c_t^\dagger = c_t$, and $Q_t^\dagger = 0$, and also with initial conditions $a_1^\dagger = 0$ and $P_1^\dagger = P_{\infty,1} = BB'$. Notice that $P_{\infty,t}$ in (3.18) is the Kalman mean square error prediction equation [cf. Durbin and Koopman 2001, Eq. (4.11)] applied to the auxiliary model, and recall that, from Assumption 3.12, $F_{\infty,t}$ cannot be a zero matrix. Supposing first $F_{\infty,t}$ is nonsingular, take the recursive formulae of $r_{t-1}^{(0)}$ and $r_{t-1}^{(1)}$ in (3.19) and place them in the expression of $a_{t|n}$, which gives

$$a_{t|n} = a_t^{(0)} + P_{\infty,t} \tilde{Z}_t' F_{\infty,t}^{-1} v_t^{(0)} + P_{\infty,t} L_t^{(0)'} r_t^{(1)} + P_{\infty,t} L_t^{(1)'} r_t^{(0)} + P_{*,t} L_t^{(0)'} r_t^{(0)}. \tag{3.21}$$

From (3.21), identity (3.20) will be proved whenever the following three claims are established.

Claim 1 $A_t \left(a_t^{(0)} + P_{\infty,t} \tilde{Z}_t' F_{\infty,t}^{-1} v_t^{(0)} \right) = q_t.$

Proof. Define $a_{t|t}^{(0)} \equiv a_t^{(0)} + P_{\infty,t} \tilde{Z}_t' F_{\infty,t}^{-1} v_t^{(0)}$. Looking at the recursions in (3.18), it follows that $a_{t|t}^{(0)}$ is the Kalman updating equation [cf. Durbin and Koopman 2001, Eq. (4.14)] applied to the auxiliary model.   □

Claim 2 $A_t P_{\infty,t} L_t^{(0)'} = 0.$

Proof. Still considering the auxiliary model, use the expression of $L_t^{(0)}$ in (3.18) and Lemma 3.15 to obtain

$$A_t P_{\infty,t} L_t^{(0)'} = A_t P_{\infty,t} T_t' - A_t P_{\infty,t} \tilde{Z}_t' F_{\infty,t}^{-1} \tilde{Z}_t P_{\infty,t} T_t'$$

$$= A_t P_{\infty,t} T_t' - \begin{bmatrix} 0_{k \times p} & I_{k \times k} \end{bmatrix} \tilde{Z}_t P_{\infty,t} T_t'$$

$$= 0.$$

□

Claim 3 $A_t \left(P_{\infty,t} L_t^{(1)'} + P_{*,t} L_t^{(0)'} \right) = 0.$

Proof. From the expression of $L_t^{(1)'}$ in (3.18) it follows that

$$A_t P_{\infty,t} L_t^{(1)'} = -A_t P_{\infty,t} \tilde{Z}_t' F_{\infty,t}^{-1} \tilde{Z}_t P_{*,t} T_t' + A_t P_{\infty,t} \tilde{Z}_t' F_{\infty,t}^{-1} F_{*,t} F_{\infty,t}^{-1} \tilde{Z}_t P_{\infty,t} T_t'$$

$$= - \begin{bmatrix} 0_{k \times p} & I_{k \times k} \end{bmatrix} \tilde{Z}_t P_{*,t} T_t' + \begin{bmatrix} 0_{k \times p} & I_{k \times k} \end{bmatrix}$$

$$\begin{bmatrix} \tilde{Z}_t P_{*,t} \tilde{Z}_t' + \mathrm{diag}\left(H_t, 0_{k \times k}\right) \end{bmatrix} F_{\infty,t}^{-1} \tilde{Z}_t P_{\infty,t} T_t'$$

$$= -A_t P_{*,t} T_t' + A_t P_{*,t} \tilde{Z}_t' F_{\infty,t}^{-1} \tilde{Z}_t P_{\infty,t} T_t', \tag{3.22}$$

where the second equality comes from Lemma 3.15 used with the auxiliary model and from the expression of $F_{*,t}$ in (3.18) associated with the augmented model. On the other hand, the expression of $L_t^{(0)}$ implies

$$A_t P_{*,t} L_t^{(0)'} = A_t P_{*,t} T_t' - A_t P_{*,t} \tilde{Z}_t' F_{\infty,t}^{-1} \tilde{Z}_t P_{\infty,t} T_t'. \tag{3.23}$$

Add (3.22) and (3.23).   □

If $F_{\infty,t}$ is singular, then uncouple the augmented measurement $(Y_t', q_t')'$ in such a way that

$$Y_{t,1}, \ldots, Y_{t,p-1}, (Y_{t,p}, q_t')'.$$

Since the new $F_{\infty,t}$, associated with $(Y_{t,p}, q_t')'$, is nonsingular (cf. Lemma 3.14), proceed exactly as previously to obtain (3.20).

Take now an arbitrary $t > d$. In such a case, the recursions in (3.19) convert to de Jong's version of the Kalman smoother [cf. de Jong 1989; Durbin and Koopman 2001, Eq. (4.26)] and, therefore, (3.21) assumes the simpler form

$$a_{t|n} = a_{t|t-1} + (P_{t|t-1}T_t' - P_{t|t-1}\tilde{Z}_t'F_t^{-1}\tilde{Z}_t P_{t|t-1}T_t')r_t,$$

where $r_{t-1} = \tilde{Z}_t'F_t^{-1}v_t + (T_t - T_t P_{t|t-1}\tilde{Z}_t'F_t^{-1}\tilde{Z}_t)' r_t$, with $r_n = 0$. Using almost the same arguments, prove slight modifications of Claims 1 and 3 to obtain identity (3.20); Claim 2 is trivially satisfied, given that $P_{\infty,t} = 0$ in this case. This completes the proof. $\qquad \square$

From a practical perspective, a point coming from this last result that must be reinforced is that, under quite general conditions, it is *always possible* to yield restricted smoothed state vectors even when the estimation lies in the diffuse period (that is, for $t = 1, \ldots d$, whatever d may be). Stated in other words: the beginning of the series is not critical anyway to obtaining more interpretable results (which is certainly the case whenever estimated state vectors under meaningful restrictions are achieved).

In Pizzinga (2012), the issue of the statistical efficiency of this initial exact restricted Kalman smoother is dealt with in terms of conditional expectation theory. The main result is that, under some specific σ-fields (or *information sets*), *conditional* mean square errors associated with state vector smoothing – even in the diffuse period $\mathcal{D}$ – are always improved whenever one considers an augmented model suggested in Theorem 3.1.

Chapter 4
Restricted Kalman Filtering: Methodological Issues

This chapter is concerned with some methods for imposing linear restrictions in state space modeling. The plan I will follow is this. In Sect. 4.1, I discuss an alternative restricted Kalman filtering that is indicated for situations where the linear restrictions are time-invariant and the state vector follows a general random walk. This approach was first featured in Pizzinga (2009). In Sect. 4.2, I present another alternative restricted Kalman filtering, this time due to Pizzinga (2010), that is based on a reduced linear state space model; this method will be compared with the previous augmented restricted Kalman filtering from several standpoints. Finally, Sect. 4.3 deals with a method developed in Pizzinga (2010) to impose linear restrictions in the prediction of a state vector.

4.1 Random-Walk State Vectors Under Time-Invariant Restrictions

In this section, the paradigm of augmenting the measurement equation, in order to accomplish linear restrictions in state vector estimation, changes. Actually, this brief change in course deserves some attention because it may highlight a potential framework in restricted Kalman filtering.

The result of this section, whose proof is still carried out by elementary Hilbert space theory, is as follows.

Theorem 4.1. *If the linear state space model in (2.1) is such that $c_t = 0$ and $T_t = R_t = I$, then (i) $A\alpha_1 = q$ (with q deterministic) and (ii) $AQ_t A' = 0$ for all t=1,2,... are sufficient to yield*

$$Aa_{t|j} = q \quad for \ all \ t, j = 1, 2. \ldots \tag{4.1}$$

A. Pizzinga, *Restricted Kalman Filtering: Theory, Methods, and Application*, SpringerBriefs in Statistics 12, DOI 10.1007/978-1-4614-4738-2_4, © Springer Science+Business Media New York 2012

Proof. Fix t and j. Once again, denote by $\pi_{S'}$ the linear orthogonal projection onto S'. Now observe that, from a trivial recursion on the state equation,

$$\alpha_t = \alpha_1 + \sum_{j=1}^{t-1} \eta_{t-j}. \tag{4.2}$$

Premultiplying both sides of (4.2) by A implies

$$A\alpha_t = A\alpha_1 + \sum_{j=1}^{t-1} A\eta_{t-j} = q + 0 = q, \tag{4.3}$$

where the second equality comes from hypotheses (i) and (ii). Denoting the ith row from A by $A_i = [c_{i1} \ldots c_{im}]$, it follows that

$$A_i a_{t|j} = c_{i1} a_{t1|j} + \cdots + c_{im} a_{tm|j} = c_{i1} \pi_{S'}(\alpha_{t1}) + \cdots + c_{im} \pi_{S'}(\alpha_{tm})$$

$$= \pi_{S'}(c_{i1}\alpha_{t1} + \cdots + c_{im}\alpha_{tm}) = \pi_{S'}(A_i \alpha_t) = \pi_{S'}(q_i)$$

$$= q_i,$$

where the third, fifth, and sixth equalities come respectively from the linearity of $\pi_{S'}$, from (4.3), and from the fact that $q_i \in \mathcal{R}(\pi_{S'}) = S'$. Since t, j, and i were taken arbitrarily, the theorem is proved. $\square$

I should make explicit some practical gains from this last proposition – which was given and proved in Pizzinga (2009) – applicable to models in which the state vector evolves as (possibly heteroscedastic) random walks. The first bonus is that there is no longer any need to increase the dimension of the measurement equation. The second plus is that, by imposing the enunciated restrictions on the initial state vector and on the covariance matrices of the error terms from the state equation, maximum likelihood estimation can be sharply enhanced whenever some of the unknown parameters belong to those matrices. The third advantage is that the restrictions are satisfied by any type of state estimation, whether it is a prediction, updating, or smoothing.

4.2   Reduced Restricted Kalman Filtering

4.2.1   Motivation

In dealing with a linear regression model under linear restrictions, there are two ways to estimate. Actually, both prove to be numerically equivalent and are known by the name of restricted least squares. The first way was already revisited in Sect. 3.3 [cf. expressions in (3.15)], while the second is implemented by rewriting a reduced model with transformed data and then applying the usual OLS estimation to the transformed data (cf. Davidson and Mackinnon 1993).

Following Pizzinga (2010), this section presents a restricted Kalman filtering under a reduced modeling framework. If the usual restricted Kalman filtering by augmentation discussed so far can be viewed as a generalization of the first way to impose linear restrictions in a static linear regression model, the approach to be discussed now, in turn, resembles the second. One feature to be listed among others is that, even though both approaches to restricted least squares produce exactly the same result, the two restricted Kalman filtering (the augmented and the reduced) *do not* always result in the same estimated state vectors.

4.2.2 The Method

In the remainder of this Sect. 4.2, consider the measurement equation in (2.1), the restrictions in (2.5), and the following assumption.

Assumption 4.2. *The (possibly random) vector* $q_t = (q_{t1}, \ldots, q_{tk})'$ *is such that* $q_{ti} \in span\{1, Y_{11}, \ldots, Y_{1p}, \ldots, Y_{t1}, \ldots, Y_{tp}\}$, *for all* $i = 1, \ldots, k$ *and* $t = 1, 2, \ldots$.

The basic perspective behind the alternative restricted Kalman filtering is much the same as that of the reduced modeling in linear regression under linear restrictions: some state coordinates are rewritten as affine functions of the others and the result is appropriately placed in the measurement equation.

The *method*:

Let t be an arbitrary time index.

1. Without loss of generality write the linear restrictions in (2.5) as

$$A_{t,1}\alpha_{t,1} + A_{t,2}\alpha_{t,2} = [A_{t,1} \ A_{t,2}] \left(\alpha'_{t,1}, \alpha'_{t,2}\right)' = q_t, \quad (4.4)$$

   where $A_{t,1}$ is a $k \times k$ full rank matrix.

2. Solve (4.4) for $\alpha_{t,1}$, which should result in

$$\alpha_{t,1} = A_{t,1}^{-1} q_t - A_{t,1}^{-1} A_{t,2}\alpha_{t,2}. \quad (4.5)$$

3. Take (4.5) and put it in the measurement equation of the model in (2.1) – from which d_t is dropped without loss of generality – aiming to obtain

$$
\begin{aligned}
Y_t &= Z_{t,1}\alpha_{t,1} + Z_{t,2}\alpha_{t,2} + \varepsilon_t \\
&= Z_{t,1}\left(A_{t,1}^{-1} q_t - A_{t,1}^{-1} A_{t,2}\alpha_{t,2}\right) + Z_{t,2}\alpha_{t,2} + \varepsilon_t \\
&= Z_{t,1}A_{t,1}^{-1} q_t - Z_{t,1}A_{t,1}^{-1} A_{t,2}\alpha_{t,2} + Z_{t,2}\alpha_{t,2} + \varepsilon_t \\
\Rightarrow Y_t^* &\equiv Y_t - Z_{t,1}A_{t,1}^{-1} q_t = \left(Z_{t,2} - Z_{t,1}A_{t,1}^{-1} A_{t,2}\right)\alpha_{t,2} + \varepsilon_t \\
&\equiv Z_{t,1}^*\alpha_{t,2} + \varepsilon_t.
\end{aligned}
$$

4. Postulate a transition equation for the unrestricted state vector $\alpha_{t,2}$. This equation leads to the following *reduced* linear state space model:

$$Y_t^* = Z_{t,2}^* \alpha_{t,2} + \varepsilon_t \ , \ \ \varepsilon_t \sim \text{WN}(0, H_t),$$
$$\alpha_{t+1,2} = T_{t,2}\alpha_{t,2} + c_{t,2} + R_{t,2}\eta_{t,2} \ , \ \ \eta_{t,2} \sim \text{WN}(0, Q_{t,2}), \tag{4.6}$$

where $E(\alpha_{1,2}) = a_{1,2}$ and $Var(\alpha_{1,2}) = P_{1,2}$.

5. For the reduced model in (4.6), apply the usual Kalman filtering to obtain $a_{t,2|j}$ and $P_{t,2|j}$, for all $j \geq t$.

6. Reconstitute the estimate $a_{t,1|j}$ and its mean square error matrix $P_{t,1|j}$ by means of the affine relation given in (4.5):

$$a_{t,1|j} = A_{t,1}^{-1} q_t - A_{t,1}^{-1} A_{t,2} a_{t,2|j},$$
$$P_{t,1|j} = (A_{t,1}^{-1} A_{t,2}) P_{t,2|j} (A_{t,1}^{-1} A_{t,2})'. \tag{4.7}$$

Some comments on this algorithm are in order. First, the approach, which is taken from Pizzinga (2010) under the same notations, was first considered by Leybourne (1993) and by Doran and Rambaldi (1997); what Pizzinga (2010) did was to put it in a more general framework. In addition, observe that j *does have to be greater than or equal to t* due to steps 5 and 6 (cf. Assumption 4.2). Another aspect is that the specification for the state equation in step 4 could be extracted from the *complete* state equation in (2.1), but if one does not want to think or worry about a full transition system, then one could concentrate only on modeling the block $\alpha_{t,2}$.

4.2.3   Reducing Versus Augmenting

As cited in Pizzinga (2010), there are several advantages of the reduced model approach over the augmented model:

- *Mathematical consistency.* Once the state equation is chosen after the reducing task, the method avoids any risk of obtaining measurement and state equations theoretically inconsistent with each other.
- *Computational efficiency.* While the augmenting approach increases the dimension of the practical problem (the length of the measurement vectors increases from p to $p + k$!), the reduced model approach goes in an opposite direction by not altering the size of the measurement equation and shortening the size of the state equation (from m to $m - k$). In other words, the augmenting approach "augments" the dimensions of the practical problem while the reduced model approach "reduces" them.
- *Model selection.* The reduced model approach enables one to investigate the plausibility of the assumed linear restrictions by using information criteria (e.g., AIC and BIC). The competing model would be the unrestricted one as given by (2.1), the (*quasi-*)likelihood function of which is surely comparable with that from the restricted model in Eq. (4.6).

Stepping further toward the comparison between the reducing and the augmenting approaches, two results are demonstrated in Pizzinga (2010). Both are related to the augmented model suggested in Theorem 3.1 and reveal that, for certain types of state restrictions, that model is much less flexible. The first proposition concerns limitations on the state equation. Note that the first three conditions listed below are quite general since they are verified for several state space specifications (e.g., zero-mean initial state vectors, whether diffuse or nondiffuse) and for many types of linear restrictions (e.g., all the deterministic ones).

Proposition 4.3. *Suppose the partition in (4.4) is such that* $A_{t,1} \equiv A_1$. *Also, admit the following conditions:*

(i) $T_t = diag\,(T_{t,1}, T_{t,2})$, *where* $T_{t,1}$ *is* $k \times k$.

(ii) $\left(A_1^{-1} A_{t+1,2} T_{t,2} - T_{t,1} A_1^{-1} A_{t,2}\right) E\,(\alpha_{t,2}) + \left[I_{k\times k} \quad A_1^{-1} A_{t+1,2}\right] c_t = 0.$

(iii) $E\,(q_t) = E\,(q_{t+1}) = \bar{q}$.
   Then (i), (ii), and (iii) are sufficient for $T_{t,1} = I_{k\times k}$. *Now, suppose (i), (ii), and (iii) are valid for all* $t \geq 1$ *and consider the following additional conditions:*

(iv) $\forall t \geq 1 : A_{t,2} \equiv A_2$, *such that* A_2 *has null kernel.*

(v) $\forall t \geq 1 : q_t \equiv q$ *(possibly random).*
   Now, (i)–(v) are sufficient for $T_t = I_{m\times m}$.

It becomes clear from Proposition 4.3 that, if one chooses the augmenting approach for dealing with important types of restrictions, there would be no possibility left but a random-walk evolution for at least a block of the state vector.

The second proposition is stated below. Its condition (vii), as one can directly see, is a quite natural setup since this avoids some pathological behaviors from the measurement equation, such as nonergodic stationarity:

Proposition 4.4. *Suppose conditions (i)–(iii) of Proposition 4.3 are valid for all* $t \geq 1$, *as well as (iv) and (v), with q deterministic. Also assume that*

(vi) $\forall t \geq 1 : Q_t \equiv diag\,(\sigma_{t1}^2, \ldots, \sigma_{tm}^2)$ *and* $R_t = I_{m\times m}$;

(vii) $\forall t \geq 1$ *and* $\forall i = 1, \ldots, m : \sigma_{1i}^2 = \cdots = \sigma_{ti}^2 = 0 \Rightarrow Var\,(\alpha_{1i}) = 0.$

Then, $Q_t = O_{m\times m}$ *for all* $t \geq 1$.

This last result rules out any possibility of nondegenerated state vectors under contemporaneously uncorrelated errors $\eta_{t1}, \ldots, \eta_{tm}$. This limitation, as that previously raised from Proposition 4.3, surely does not arise under the reducing approach.

A final comment about this reducing approach is that it also has a strong geometrical appeal, under which it becomes possible to interpret such a method as a "two-stage" state vector estimation – or a "partitioned" projection of the state vector. For more details, the reader is referred to Pizzinga (2010).

4.3   Predictions from a Restricted State Space Model

The original proposal for adopting an augmented model, which I reevoke again in this section, does not, in general, guarantee that linear restrictions on the state vector will be carried over to the Kalman prediction equations (*immediate example*: there is no extension of Corollary 3.2 for the prediction equations when one is dealing with any of the structural models – cf. Harvey (1989), Chaps. 2 and 4 – put in their respective state space forms).

In this section I present a strategy to further extend (that is, for *any* type of linear state space model) restricted Kalman filtering and smoothing up to schemes aimed at prediction. The approach was previously presented in Pizzinga (2010) and is built on the ideas of missing values state space treatment and of the decomposition used in the second proof of Theorem 3.1 and in the proof of Theorem 3.7.

Consider that one is willing to extrapolate the state vector or the measurements up to h steps ahead in the future; that is, one wants to obtain $a_{n+1|n}, \ldots, a_{n+h|n}$ or $\hat{Y}_{n+1|n}, \ldots, \hat{Y}_{n+h|n}$. But, like everything that has been done so far in this book, it is known *a priori* that, for all $j = 1, \ldots, h$, $A_{n+j}\alpha_{n+j} = q_{n+j}$, where A_{n+j} is a $k \times m$ matrix and q_{t+j} is a $k \times 1$ (possibly random) vector; this knowledge is simply the confirmation that Assumption 2.1 is not confined to a particular time series of size n. So the question is how to make $a_{n+1|n}, \ldots, a_{n+h|n}$ satisfy those same theoretical constraints.

Everything starts again with the adoption of an augmented model. The augmented version of (2.1) is rewritten to accomplish the "future" restrictions:

$$\begin{pmatrix} Y_t \\ q_t \end{pmatrix} = \begin{pmatrix} Z_t \\ A_t \end{pmatrix} \alpha_t + \begin{pmatrix} d_t \\ 0 \end{pmatrix} + \begin{pmatrix} \varepsilon_t \\ 0 \end{pmatrix}, \quad \begin{pmatrix} \varepsilon_t \\ 0 \end{pmatrix} \sim \mathrm{WN}\left(\begin{pmatrix} 0 \\ 0 \end{pmatrix}, \begin{pmatrix} H_t & 0 \\ 0 & 0 \end{pmatrix} \right) \quad (4.8)$$

$$\alpha_{t+1} = T_t \alpha_t + c_t + R_t \eta_t \ , \quad \eta_t \sim \mathrm{WN}(0, Q_t),$$

$$t = 1, \ldots, n, n+1, \ldots, n+h.$$

Moving ahead, observe now that the model in (4.8) can be decomposed in a way that stresses that one is actually dealing with the (possibly multivariate) series

$$Y_1, q_1, Y_2, q_2, \ldots, Y_n, q_n, Y_{n+1}, q_{n+1}, Y_{n+2}, q_{n+2}, \ldots, Y_{n+h}, q_{n+h}, \quad (4.9)$$

which has missing measurements; $Y_{n+1}, Y_{n+2}, \ldots, Y_{n+h}$ are obviously absent. So the series in (4.9) presents blanks and should be appropriately recast as

$$Y_1, q_1, Y_2, q_2, \ldots, Y_n, q_n, \quad , q_{n+1}, \quad , q_{n+2}, \ldots, \quad , q_{n+h}. \quad (4.10)$$

The obtention of $a_{n+1|n}, a_{n+2|n}, \ldots, a_{n+h|n}$ is then almost equivalent to the application of Kalman smoothing to the "incomplete" series in (4.10) using the following equivalent version of (4.8). The measurement equation is defined by

$$Y_{t,i} = Z_{t,i}\alpha_{t,i} + d_{t,i} + \varepsilon_{t,i} \ , \ \varepsilon_{t,i} \sim \text{WN}(0, H_{t,i}).$$

When $i = 1$, nothing is changed from the measurement equation in (2.1) of Sect. 2.1. But for $i = 2$ we must have

$$Y_{t,2} = q_t, \quad Z_{t,2} = A_t, \quad d_{t,2} = 0 \text{ and } H_{t,2} = 0.$$

Regarding the state equation, for all t, $\alpha_{t,2} = \alpha_{t,1}$, and $\alpha_{t+1,1} = T_t\alpha_{t,2} + c_t + R_t\eta_t$, $\eta_t \sim \text{WN}(0, Q_t)$. The use of the word "almost" is justified by the fact that the information from $q_{n+1}, \ldots, q_{n+h}$ does enter into the Kalman estimation; so the result does not necessarily equal $a_{n+1|n}, a_{n+2|n}, \ldots, a_{n+h|n}$, which theoretically use the information only up to q_n. The modifications in the original Kalman equations due to missing observations are discussed in Durbin and Koopman (2001), Sect. 4.8. Finally, notice that, from Theorem 3.1, it follows that, for all $j = 1, \ldots, h$, $A_{n+j}a_{n+j|n} = q_{n+j}$. For the reasons just explained in this paragraph, "$n + j|n$" is an abuse of notation.

The earlier method can be gathered into the following algorithm:

1. Decompose the model in (4.8) aiming to get the series in (4.10).
2. Store the "new" observations while respecting the missing-value positions.
3. Apply the Kalman smoothing equation to the stored observations, appropriately modified to account for the missing values.
4. Take the smoothed states corresponding to the missing-value positions as the *predicted state vectors under linear restrictions*.

Chapter 5
Applications

Following the presentation and discussion of several theoretical and methodological issues of previous chapters, this chapter will be devoted entirely to some practical examples, where the two methods from Chap. 4 – reduced restricted Kalman filtering and the restricted Kalman prediction equations – are considered, implemented, and evaluated. The remainder of this chapter is structured as follows. Section 5.1 presents an application, previously conducted in Pizzinga et al. (2011), where a time-varying factor modeling for dynamic style analysis is implemented. In this application, an accounting restriction on the coefficients is tackled by reduced restricted Kalman filtering. In Sect. 5.2, the time-varying econometric models proposed in Souza et al. (2011) are considered for the estimation and interpretation of the dynamic exchange-rate pass-through over Brazilian price indexes; again, reduced restricted Kalman filtering is key to accessing some economic hypotheses imposed under two specific restrictions. And, in Sect. 5.3, the material concerning restricted predictions from Sect. 4.3 is conveniently implemented for obtaining predictions of quarterly gross domestic product (GDP) that must be somehow consistent with the annual GDP (that is, for each year, the sum of quarterly GDP data *must* equal the annual GDP), much in the same way as was done in one of the applications conducted in Pizzinga (2010).

The models to be discussed in the sequel were implemented using the Ox 3.0 language (cf. Doornick 2001) with occasional use of the Ssfpack 3.0 library for linear state space modeling (Koopman et al. 2002). The implementations were carried out on an Athlon XP 2200 MHz, with 378 MB RAM. The computational efficiency of the estimations are separately analyzed and discussed in the appropriate sections. All the estimations were done under the *exact maximum-likelihood estimation* and the *exact initial Kalman filter* (cf. Durbin and Koopman 2001, Chaps. 5 and 7).

A. Pizzinga, *Restricted Kalman Filtering: Theory, Methods, and Application*,
SpringerBriefs in Statistics 12, DOI 10.1007/978-1-4614-4738-2_5,
© Springer Science+Business Media New York 2012

5.1   Case I: Semistrong Dynamic Style Analysis

5.1.1   Motivation

Depending on the type of investment fund under investigation, detailed information on the actual portfolio composition is not usually available. *Return-based
style analysis*, or simply *style analysis*, is a statistical method for the estima-
tion/approximation of the unknown composition of an investment fund portfolio.
The standard practice in style analysis only uses the so-called *external information*,
which is represented by the fund returns and some market index returns, and is
implemented by the *asset class factor model* (cf. Sharpe 1988, 1992). Later, this was
modified by the addition of an intercept term (cf. de Roon et al. 2004), as follows:

$$R_t^P = \alpha + \beta_1 R_{t1} + \beta_2 R_{t2} + \cdots + \beta_m R_{tm} + \varepsilon_t. \tag{5.1}$$

Assumptions: The process R_t^P is the portfolio return. The m-variate process $R_t =$
$(R_{t1}, R_{t2}, \ldots, R_{tm})'$ represents some asset class index returns, which should satisfy
the assumptions of *exhaustiveness*, *mutual exclusiveness*, and *different behavior*
(cf. Sharpe 1988, 1992). The coefficients $\beta_1, \beta_2, \ldots, \beta_m$ are the unknown alloca-
tions/exposures that are sometimes supposed to satisfy an accounting constraint
known as the *portfolio restriction*, that is, $\sum_{i=1}^m \beta_i = 1$. There is also a *short-
sale restriction*, which is sometimes considered and is implemented by forcing the
nonnegativeness of $\beta_1, \beta_2, \ldots, \beta_m$. But, as this restriction is not always meaningful
(e.g., most hedge funds take positions in derivative markets), this practice is not
adopted here. The intercept α is the *Jensen measure* or *Jensen alpha* (cf. Jensen
1968; Carhart 1997; de Roon et al. 2004; and Elton et al. 2006) and represents the
idiosyncratic fund return, i.e., it measures how much the fund gains – or loses – by
means of its selectivity strategies. Finally, ε_t is a typical random error process with
finite second moments.

Although it is a widely used tool in investment analysis, model (5.1) has a
drawback: it ignores the fact that asset class exposures and selectivity *do change*
over time, reflecting the very plausible and possible reallocations of the assets by the
portfolio manager – an idea that was also evoked in Pizzinga and Fernandes (2006)
and in Swinkels and Van der Sluis (2006). Later, in Pizzinga et al. (2008b), a class
of *semistrong* style analysis models – meaning that only the portfolio restriction is
imposed; cf. the style analysis taxonomy proposed by de Roon et al. (2004) – was
proposed whose exposures and Jensen's measure were both made stochastically
time-varying as a (vector) random walk. This represented a direct generalization
of the static model (5.1). Empirical illustrations were presented using return series
of Brazilian real/US dollar exchange-rate funds. Among several points, there was
clear visual evidence that the time-varying exposures to US dollar/real exchange-
rate markets behaved under different autoregressive regimes, one of those directly
associated to the 2002 Brazilian presidential election, a period of some political
turbulence and high volatility.

This section's exercise is mostly based on the results of Pizzinga et al. (2011). First, evidence on switching regimes for the time-varying exposures of Brazilian real/US dollar exchange-rate funds is revealed and, second, an interpretation of the estimated exposures from the "more appropriate" model is given. The dynamics to be evaluated for the time-varying restricted exposures are as follows: (1) random walk, (2) simply autoregressive, (3) autoregressive with abrupt switching regimes, and (4) nonlinear under a general smoothing transition function. The elected regime-switching variable for the third and fourth models is the AR(1)-GARCH(1,1) volatility of the US dollar/real exchange rate.

5.1.2 Competing Models

I now present the analytical expressions of several *time-varying asset class factor models* for *semistrong dynamic style analysis*. In what follows, the reducing method from Sect. 4.2.2 has been evoked to make the portfolio restriction attainable.

Let us first obtain the expression corresponding to the portfolio restriction on the state vector, which I will denote in this section by γ_t and whose coordinates represent the exposures and Jensen's measure. To do this, we use steps 1 and 2 of the algorithm from Sect. 4.2.2:

$$1 = [1\ 1\ldots 1\ 0]\,(\beta_{t1},\beta_{t2},\ldots,\beta_{tm},\alpha_t)'$$

$$\Rightarrow\quad 1 = \beta_{t1} + [1\ldots 1\ 0]\,(\beta_{t2},\ldots,\beta_{tm},\alpha_t)'$$

$$\Rightarrow\ \beta_{t1} = 1 - [1\ldots 1\ 0]\,(\beta_{t2},\ldots,\beta_{tm},\alpha_t)'$$

$$\Rightarrow\ \gamma_{t,1} = 1 - [1\ldots 1\ 0]\,\gamma_{t,2}.$$

We now move on to the measurement equation of the reduced model by making use of step 3 of the algorithm in conjunction with the last equality obtained previously:

$$R_t^c = R_{t1}\beta_{t,1} + [R_{t2}\ldots R_{tm}\ 1]\,(\beta_{t2},\ldots,\beta_{tm},\alpha_t)' + \varepsilon_t$$

$$= R_{t1} - R_{t1}\,[1\ldots 1\ 0]\,(\beta_{t2},\ldots,\beta_{tm},\alpha_t)'$$

$$+ [R_{t2}\ldots R_{tm}\ 1]\,(\beta_{t2},\ldots,\beta_{tm},\alpha_t)' + \varepsilon_t$$

$$\Rightarrow R_t^c - R_{t1} = [R_{t2} - R_{t1}\ldots R_{tm} - R_{t1}\ 1]\,(\beta_{t2},\ldots,\beta_{tm},\alpha_t)' + \varepsilon_t$$

$$\Rightarrow R_t^c - R_{t1} = [R_{t2} - R_{t1}\ldots R_{tm} - R_{t1}\ 1]\,\gamma_{t,2} + \varepsilon_t.$$

Finally, combining a rather encompassing state equation with the preceding expression, we arrive at the following general structure:

$$R_t^P - R_{t,1} = [R_{t2} - R_{t1} \ldots R_{tm} - R_{t1} \quad 1] \gamma_{t,2} + \varepsilon_t, \quad \varepsilon_t \sim \mathrm{NID}(0, \sigma^2 X_t)$$
$$\gamma_{t+1,2} = diag\left(T_t^\beta, 1\right) \gamma_{t,2} + \eta_t \ , \quad \eta_t \sim \mathrm{NID}(0, Q) \tag{5.2}$$
$$\gamma_{t,1} = 1 - [1 \ldots 1 \ 0] \gamma_{t,2}.$$

We enumerate some features of model (5.2). Firstly, it should be reinforced that the last coordinate of γ_t is the time-varying Jensen measure α_t and the remaining coordinates are the time-varying exposures $\beta_{t1}, \beta_{t2}, \ldots, \beta_{tm}$. Secondly, as can be directly seen from the second line of (5.2), the Jensen measure follows a random walk. And thirdly, X_t in the measurement error's variance is some nonnegative variable that must respond to occasional heteroscedastic behavior, and Q can be set full. Pizzinga et al. (2011) defend the idea that the impacts on the exposures, represented by the components of η_t, "*do 'communicate' among themselves – it would be unreasonable to assume that investment decisions (and hence the exposures) are related only by the portfolio restriction (which is an accounting constraint), since they reflect the same underlying shocks.*"

The decision toward the random walk for the evolution of the Jensen measure was justified in Pizzinga et al. (2011) as follows. Although such choice seems too simple and perhaps "unrealistic" at first glance. Three reasons support such a judgment. The first is that of *parsimony* and *simplicity*, as there is no additional clue to guide one in choosing a more complex dynamic. The second is the allowance for the possibility of fundamental selectivity changes over time due to nonstationarity. The third is almost surely nonexplosiveness since for "large" series the smoothed Jensen measure must intercept the time x-line infinitely often with probability 1 (cf. Chung 2001, Chap. 8).

The remaining part of model (5.2)'s specification lies on the transition submatrix $T_t^\beta \equiv diag(\phi_{t2}, \ldots, \phi_{tm})$, which drives the evolution of the unrestricted block of time-varying exposures in $\gamma_{t,2}$. Let us first enumerate the possibilities to be investigated and, in the sequel, give appropriate rationalities to each of them:

1. *Random walk* (*RW*): $T_t^\beta \equiv I_{(m-1)\times(m-1)}$.
2. *Purely autoregressive* (*AR*): $T_t^\beta \equiv diag(\phi_2, \ldots, \phi_m)$, where $|\phi_i| < 1$ for all i.
3. *Autoregressive with abrupt switching regimes*: some diagonal entries of T_t^β take the form $\phi_{i1} + \phi_{i2}d_{ti}$, where $d_{ti} = 1$ if some exogenous variable z_t assumes certain values and $d_{ti} = 0$ otherwise.
4. *Nonlinear under a general smoothing transition function*: some diagonal entries of T_t^β take the form $\phi_{i,1} + \phi_{i,2}z_t + \phi_{i,3}z_t^2$, where z_t is some exogenous variable.

The first model is clearly the most parsimonious and has already been used by Pizzinga et al. (2008b) within this same style analysis framework. According to Swinkels and Van der Sluis (2006), this specification should be used if one believes that exposures can increase or decrease over time when responding to shocks (that turn out to exert a permanent effect). In contrast, if one believes that exposures can deviate for some time from normal (or "steady-state") levels but will forcefully come back to them (which means that shocks exert a transitory effect), then a variant like the second model should be used. As pointed out by Pizzinga et al. (2008b), two Brazilian real/US dollar exchange-rate funds had their time-varying Jensen measures and exposures analyzed. As was already indicated, exposures estimated under this framework seem to follow two different patterns, one during the months near the 2002 Brazilian presidential election (when exposures to US dollar/real markets appeared to be more erratic and less persistent), and the other during the remaining months (in which exposures were much more stable). This "stylized fact" was interpreted as a suggestion that more sophisticated dynamics (especially regime-switching models) should be considered.

The second model captures situations in which managers try to target "steady-state" exposures. When compared with the first model, the number of parameters grows by $m - 1$ autoregressive coefficients. Since the eigenvalues of T_t^β have absolute values strictly smaller than 1 (one), nonstationarity or "explosive" behaviors for the exposures are ruled out, which is something that brings some inferential attractiveness. In addition, this second model can be understood as a bridge to the last models.

The third and fourth models undoubtedly add complexity to the process of parameter estimation but are justified by their ability to capture the state-dependent behavior of managers and investors (which generates the aforementioned possibility of multiple regimes in exposure dynamics). One might recall that the nonlinear processes used here are respectively the threshold autoregressive (TAR) model and a general smoothing transition autoregressive (STAR) model, in which the second-order polynomial on z_t is an attempt to approximate a more general "smooth" transition function. For a comprehensive treatment of these types of regime-switching proposals outside the state space framework, see Enders (2004). Carefully note that, even though they are *nonlinear processes* (i.e., there is no corresponding Gaussian nor i.i.d. Wold decomposition – cf. Brockwell and Davis 1991, 2003), these choices for the state equation still provide us with a Gaussian "linear" space model. Once these dynamics are postulated to the state equation, parameter estimation can be accomplished under the usual paradigm of maximizing the prediction error decomposition form of the likelihood (see, for example, Harvey 1989, Chap. 3; and Durbin and Koopman 2001, Chap. 7).

5.1.3   Model Selection

As there are four ways to describe the time-varying exposures, we must discuss how to decide in practice which model seems to be the most appropriate one. We adopt the following selection mechanism:

- Likelihood-ratio (LR) tests to validate or refuse the nonlinear proposals 3 and 4.
- Information criteria such as AIC and BIC.
- Predictive power by comparing pseudo-R^2 and MSE measures.
- Diagnostic tests over the standardized innovations.

The listed strategies were fully discussed in Harvey (1989), Chap. 5. Here, the null for the LR test will be H_0: *"The parameters associated to the switching regimes are all zero."* Consequently, our test aims at comparing the "reduced" model 2 with the "complete" models 3 and 4. There is strong theoretical evidence that, asymptotically, LR $= 2\left[\log L_{Max,Comp} - \log L_{Max,Red}\right] \sim \chi_k^2$, where k is the number of parameters set to zero under the null, since at least the reduced model maintains the standards for good properties of maximum-likelihood estimation (cf. Pagan 1980).

5.1.4 Empirical Results

The asset class indexes were the CDI (the average rate charged in overnight transactions between depository institutions), the US dollar/real exchange rate (in percentage points), and observed variations in two financial indicators, Quantum Cambial and Quantum Fixed Income. The data comprise 209 observations on weekly returns from 2001 to 2004 and were obtained from *Quantum Axis* (www.quantumfundos.com.br). Two US dollar/real exchange-rate funds inside the Brazilian industry were considered: HSBC Cambial FIF and Itau Matrix US Hedge FIF. Since the fund Itau Matrix US Hedge FIF was bought by another fund – Itau B Cambial FI – at the end of 2004, the corresponding estimation was implemented with data up to the first week of November 2003 (149 observations). Additional information can be obtained at the National Association of Investment Banks (www.anbid.com).

The covariance matrix Q [cf. the state equation in (5.2)] was considered full in all the estimations. The X_t heteroscedastic variable [cf. the measurement equation of (5.2)] was chosen to be the US dollar/real AR(1)-GARCH(1,1) volatility in the analyzed period, and its standardized version was used as the z_t switching regime variable for the exposures to the US dollar/real exchange rate and the Quantum Cambial. The dummy variable d_t from the TAR specification takes 1 whenever $z_t \geq 1.3$, and 0 otherwise. This calibration was chosen to capture the period of high volatility, which took place from the last week of September 2002 (located around the 90th observation) to the third week of February 2003 (located around the 110th observation). An estimate of the threshold value could have been attempted, but that would have demanded more periods of high volatility in the data. Figure 5.1, which helps in recognizing these patterns, also illustrates what happened throughout the period. Note that the second half of 2002 was marked by a confidence crisis that surged on the eve of the Brazilian presidential elections. This crisis found very fertile ground in which to grow due to fears about the macroeconomic policies that could follow the election of the candidate leading in the polls, Luis Inacio Lula da Silva. When agents perceived that a Lula administration would not change economic fundamentals like the floating exchange rate and inflation-targeting regimes, expectations about future economic developments became more favorable, financial market indicators turned positive, and volatility dropped.

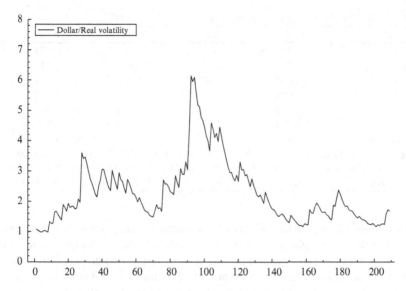

Fig. 5.1 US dollar/real volatility under the AR(1)-GARCH(1,1) model

Table 5.1 Results from estimations with HSBC FIF Cambial

Attribute	RW	AR	TAR	STAR
Log-likelihood	−149.462	−148.124	−143.421	−143.657
Computational time	0.61	3.85	6.76	10.66
Pseudo-R^2	0.905	0.907	0.905	0.905
MSE	0.549	0.534	0.545	0.550
AIC	1.516	1.561	1.554	1.576
BIC	1.756	1.801	1.858	1.912
Linearity LR test	–	–	9.406 (0.009)	8.935 (0.063)
Ljung–Box test (30 lags)	54.984 (0.004)	52.910 (0.006)	61.172 (0.001)	51.236 (0.009)
Homoscedasticity F test	0.435 (0.010)	0.481 (0.02)	0.384 (0.003)	0.442 (0.011)
Jarque–Bera test	18.694 (0.000)	18.478 (0.000)	60.247 (0.000)	11.182 (0.004)

Upon careful inspection of the information presented in Tables 5.1 and 5.2, several points emerge. Looking first at the computational efficiency, it is clear that the computation times, even though larger for the nonlinear proposals, remain essentially negligible. This could be of great value should one try to use/implement these dynamic style-analysis proposals in practice.

Moving further, one should note that the predictive power of the competing proposals gives us no clue as to which model is the most adequate – it seems that, for these particular estimations, all models can reproduce the data almost under similar capabilities (cf. pseudo-R^2 and MSE measures). Also, the use of AIC and BIC criteria is of no help in deciding which model should be considered since none of them is much larger – or smaller – than the others.

Table 5.2 Results from the estimations with Itau Matrix US Hedge FIF

Attribute	RW	AR	TAR	$STAR$
Log-likelihood	−243.201	−242.417	−234.786	−233.996
Computational time	0.33	0.77	11.1	5.44
Pseudo-R^2	0.697	0.679	0.761	0.684
MSE	3.008	3.029	2.366	3.009
AIC	3.489	3.479	3.430	3.446
BIC	3.793	3.782	3.814	3.871
Linearity LR test	–	–	15.262 (0.000)	16.841 (0.002)
Ljung–Box test (30 lags)	32.144 (0.361)	37.576 (0.161)	31.757 (0.379)	37.376 (0.166)
Homoscedasticity F test	0.670 (0.210)	0.899 (0.738)	0.944 (0.857)	0.779 (0.434)
Jarque–Bera test	6.286 (0.043)	4.242 (0.120)	7.412 (0.024)	2.618 (0.270)

The diagnostic tests in the last three lines, which were applied with the standardized innovations, though, uncover important aspects. They actually indicate that, in terms of basic model assumptions, the $STAR$ proposal systematically behaves better than the others. This is an indication that, in the analyzed period, exchange-rate exposures were driven by some switching regime nonlinear process.

Finally, looking at the results from the LR linearity test, there is evidence, at least under a 10% significance level, that the $STAR$ specification is supported by the data.

Taking into account these findings, there is no option left but to accept, among the four considered proposals, the $STAR$ model as the best description of the exposures to the exchange-rate markets.

Figures 5.2 and 5.3 depict time plots for the restricted Kalman smoothing estimates of Jensen's measure and of the exposures to US dollar/real exchange-rate spot markets and Quantum Cambial. Visual inspection suggests that the investment strategy followed by the managers of HSBC FIF Cambial is such that exposures to US dollar/real exchange-rate spot markets were negligible throughout the sample, except during the period of higher volatility, when a significant long position was taken. Furthermore, exposures to Quantum Cambial were always significant, hovering around a share of approximately 75% of the portfolio throughout the period. This outcome probably reflects preventive measures taken by fund managers during the crisis, which protected the portfolio against the losses caused by the decrease in the market value of dollar-indexed bonds issued by the Brazilian government. Managers of Itau Matrix US Hedge FIF, in turn, followed an investment strategy in which the exposures to US dollar/real exchange-rate spot markets hovered around 75 to 80% of the portfolio throughout the sample (even though the large confidence intervals observed during the period prevent ascertaining this); on the other hand, exposures to Quantum Cambial were negative and significant at several occasions.

Now, looking at the information extracted by the model on selectivity skills by analyzing the time path of Jensen's measure, the graphs at the top of Figs. 5.2 and 5.3 suggest that managers of HSBC FIF Cambial and Itau Matrix US Hedge FIF revealed a slight tendency to generate gains during the period marked by the

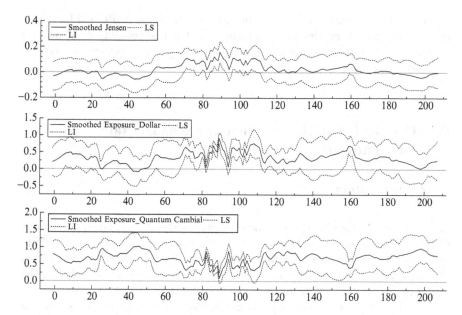

Fig. 5.2 Smoothed *STAR* exposures and Jensen's measure for HSBC FIF Cambial with respective 95% confidence intervals

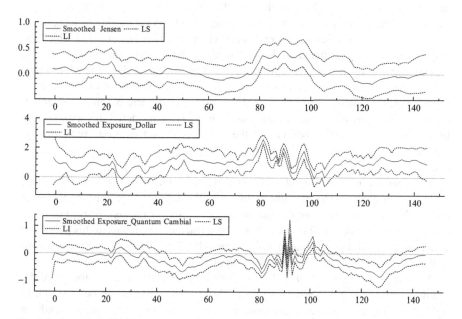

Fig. 5.3 Smoothed *STAR* exposures and Jensen's measure for Itau Matrix US Hedge FIF with respective 95% confidence intervals

confidence crisis. One can understand these facts concerning HSBC and Itau funds by recalling that it is precisely during periods of increased volatility that managers have significant profit opportunities by engaging in high-frequency operations (e.g., day-trade transactions).

5.2 Case II: Estimation of Dynamic Exchange-Rate Pass-Through

5.2.1 Motivation

In this section, linear state space models are proposed to estimate the pass-through of Brazilian price indexes against the US dollar/real exchange rate from 1996 to 2005. The methodological framework encompasses the reduced restricted Kalman filtering from Sect. 4.2, which permits one to check some economic hypotheses. The results of this section were featured originally in Souza et al. (2011).

In an open economy, domestic prices can be affected by external shocks, whether from relative currency price adjustments or from movements of international supply and demand. The exchange rate is a quite volatile economic variable in macroeconomic policy. How much does the exchange rate affect the economy? One of the faster channels is into prices. This channel is called the *exchange-rate pass-through*. Few studies have been conducted on this effect in Brazil in which the response of prices to a change in the exchange rate is suitably examined.

The importance of pass-through estimation has increased since the adoption of inflation-targeting regimes (cf. Fraga et al. 2003) and the recognition that it is crucial for inflation forecasting. In addition to these motivations, there is some evidence of a time-varying pass-through, though few studies have considered this assumption. Indeed, as Parsley (1995) points out, the stability of exchange-rate pass-through has not been well tested in common econometric specifications of pass-through equations.

The three main objectives of Souza et al. (2011), which are replicated here, are to decide whether models of null (or full) pass-through are acceptable to the price indexes investigated; to carry out likelihood ratio tests for the significance of some exogenous economic variables, which will be termed determinants in this paper and which are theoretically associated with the pass-through; and to analyze the behavior of the estimated pass-through from the best models.

According to Menon (1996), Taylor (2000), and Campa and Goldberg (1995), the main drivers of price sensitivity to exchange-rate changes can be inferred. In light of the literature with a macroeconomic approach, the pass-through depends on *inflation persistence, the degree of openness of the economy*, the *output gap*, and *real exchange-rate disalignments*. From the standpoint of disaggregated analysis, the exchange-rate pass-through is also associated with the *degree of competition of each industry* and with *firms' market power* (with the elasticity price-demand).

In light of the considerations of the last paragraph, Souza et al. (2011) proposed the following state space model for the exchange-rate pass-through for a given index price:

$$\Delta \log\, p_t = \sum_{k=1}^{m} \beta_{kt} \Delta \log\, e_{t-k} + \psi_0 + \psi_1 \Delta \log(ap_t) + \epsilon_t, \epsilon_t \sim \text{NID}(0, \sigma^2), \quad (5.3)$$

$$\beta_{t+1} = \beta_t + \gamma_1 \Delta \log(IPA_t)\mathbf{1}_{q\times1} + \gamma_2 \Delta \log(ip_t)\mathbf{1}_{q\times1}$$

$$+\gamma_3 \Delta \log(re_t)\mathbf{1}_{q\times1} + \gamma_4 \Delta \log(o_t)\mathbf{1}_{q\times1} + \xi_t, \quad \xi_t \sim \text{NID}(0, Q). \quad (5.4)$$

The first equation linearly relates the observed monthly log-variation of price to the log-variation of exchange rate until time $t - m$ and to an exogenous variable, the log-variation of the American price index, ap_t. The coefficients of $\Delta \log\, e_{t-k}$ in Eq. (5.4) are the state coordinates, which represent the *components* of the pass-through (their sum is termed *long-run pass-through*) and whose dynamics are given in Eq. (5.4), which also sets the impact from the following determinants: IPA series that represents the inflationary environment; ip_t is the industrial production index, re_t is the exchange-rate disalignment, and o_t is the openness of the economy. The matrix $Q_{m\times m}$ is set diagonal, even though the components from the pass-through (i.e., the state coordinates) do maintain degrees of dependency due to the presence of common determinants in the state equation.

Reduced restricted Kalman filtering must be evoked to make the restrictions of *full* pass-through ($\sum_{i=1}^{m} \beta_{it} = 1$) and of *null* pass-through ($\sum_{i=1}^{m} \beta_{it} = 0$) attainable. The completeness of the exchange-rate pass-through (the first restriction) means that all the variation of the exchange rate is passed to the domestic prices. This is key from the standpoint of economic theory since it means that the *purchasing power parity* (PPP) hypothesis is acceptable. On the other hand, accepting that the null exchange-rate pass-through model is the most adequate scenario implies that exchange-rate movements do not have any effect on domestic prices, and so the monetary authority need not be concerned with exchange-rate movements to make monetary policy with such price indexes.

In addition to verifying the hypotheses of completeness (or absence) of exchange-rate pass-through, another purpose of this application is to identify the most adequate number of lags of the exchange rate, that is, the value of m. For this, the same steps listed in Sect. 5.1.3 will be used.

Finally, the significance of the parameters ψ_0, ψ_1, γ_1, γ_2, γ_3, and γ_4 will be tested under a likelihood ratio (LR) testing approach. Since both the reduced and the complete model maintain the standards for good properties of maximum-likelihood estimation (cf. Pagan 1980), it follows that, asymptotically, $\text{LR} \equiv 2\left[\log L_{Max,Comp} - \log L_{Max,Red}\right] \sim \chi_1^2$, in which $log L_{Max,Red}$ represents the maximum of the log-likelihood for a model with a particular explanatory variable dropped from the specification.

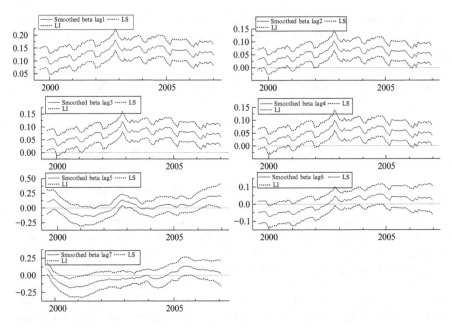

Fig. 5.4 IPA smoothed betas

5.2.2 Empirical Results

The analyzed data contain monthly observations from August 1999 to January 2007 of the Brazilian wholesale price index (IPA), the Brazilian consumer price index (IPC), the American price index, the exchange rate between the Brazilian real and the American dollar, the Brazilian industrial production index, and a measure of openness, which is the sum of imports and exports as a proportion of the GDP. The decision to use data since August 1999 is justified by the inflation-targeting system adopted by the Banco Central (counterpart to the American Federal Reserve in Brazil) in June 1999. The data were obtained from IPEA Data (www.ipeadata.gov.br), and each estimation took less than 2 s, something that highlights the computational efficiency of the adopted state space framework.

5.2.2.1 Overall IPA

The most adequate model for the IPA series is a model with seven lags on the exchange rate. Even though only the first four state vector coordinates (that is, the first four components of the pass-through) have a confidence interval that does not contain zero, the decision to retain seven lags was based on the lack of serial correlation for the residuals. Figure 5.4 shows the evolution of the coefficients over time. The pseudo-$R^2 = 0.64$ suggests that the model provides a reasonable

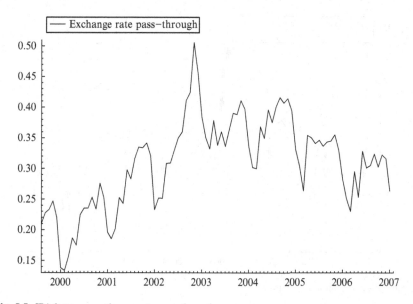

Fig. 5.5 IPA long -un exchange-rate pass-through

Table 5.3 IPA information criteria of unrestricted and restricted models

Criterion	Unrestricted	$\sum_{i=1}^{7} \beta_{i,t} = 0$	$\sum_{i=1}^{7} \beta_{i,t} = 1$
AIC	3.000	3.746	4.326
BIC	3.583	4.274	4.854

Table 5.4 IPA estimated parameters, with corresponding p-values in parentheses

ψ_1	γ_1	γ_2	γ_3	γ_4
0.001 (0.999)	0.000 (0.999)	0.001 (0.999)	0.001 (0.505)	0.000 (0.999)

adjustment for the IPA. The long-run pass-through given in Fig. 5.5 has some variation when we compare the beginning of the sample to the end with an edge at 2002, the year of elections preceding the Lula administration in Brazil, a period of great volatility in the exchange rate.

The restricted models were estimated to verify whether the hypothesis of null and full exchange-rate pass-through had support from the data. The information criteria shown in Table 5.3 provide no evidence that these extremes allow a better fit. The LR significance tests are given in Table 5.4. The p-values reveal no evidence that the proposed determinants help to explain the behavior of the pass-through.

5.2.2.2   First-Level IPA Disaggregation

To evaluate the disaggregation effects on the exchange-rate pass-through, Souza et al. (2011) considered its estimation for some groups of products. The first level of disaggregation splits the overall IPA into two main groups: consumption and production goods.

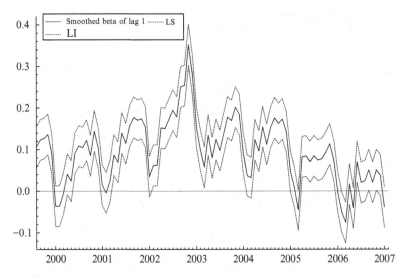

Fig. 5.6 IPA consumption-smoothed beta

Table 5.5 IPA consumption-estimated parameters, with corresponding p-values in parentheses

ψ_1	γ_1	γ_2	γ_3	γ_4
0.561 (0.000)	−0.0004 (0.000)	0.012 (0.061)	0.005 (0.006)	−0.002 (0.347)

The more adequate model for the IPA consumption series has only a lag in the exchange rate since it has the lower information criteria values and its residuals show no serial correlation. The pseudo-$R^2 = 0.62$ provides evidence in favor of goodness-of-fit. Since the decision to have only one lag for the exchange rate, the short- and long-run exchange-rate pass-through is the same. Its variation over time can be seen in Fig. 5.6. During 2002, the exchange-rate pass-through presented higher values compared to the rest of the sample period, probably due to the same explanations already given. Also, there is some indication of seasonal patterns since the pass-through seems to be close to zero at the very beginning of each year.

As shown in Table 5.5, the LR significance tests reveal that three proposed determinants are supported by the data. In addition, the inertial parameter ψ_1 is statistically significant for the measurement equation.

As happened with the IPA consumption series, the most adequate model of the IPA production series was the model with only one lag in the exchange-rate pass-through. Again, the high value of the pseudo-$R^2 = 0.729$ provides us with some confidence that the model fits the data in a proper way. The pass-through variation over time can be seen in Fig. 5.7. This indicates some aspects similar to those found in the previous analysis, except for the lack of evidence on seasonality.

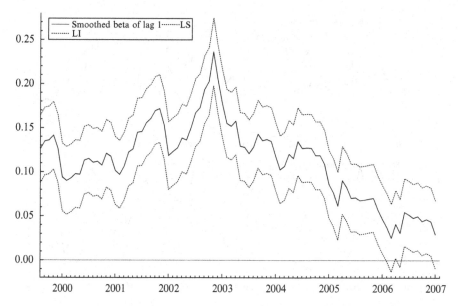

Fig. 5.7 IPA production-smoothed beta

Table 5.6 IPA production-estimated parameters, with corresponding p-values in parentheses

ψ_1	γ_1	γ_2	γ_3	γ_4
0.590 (0.000)	−0.001 (0.699)	0.002 (0.000)	0.002 (0.000)	−0.001 (0.823)

The LR significance tests shown in Table 5.6 provide us with two statistically significant determinants. Still, the inertial parameter ψ_1 is again statistically significant in the measurement equation.

5.2.2.3   IPC

The model adjusted with two lags in the exchange rate shows that the IPC does not seem to be responding to the exchange-rate movements. As can be seen in Fig. 5.8, the states corresponding to all lags vary around zero within the whole sample period. The long-run pass-through presented in Fig. 5.9 also oscillates around zero. This will be taken as the first symptom of an absence of pass-through, and this is reinforced by the application of restricted Kalman filtering since the model with a null pass-through restriction has the best information criteria (Table 5.7).

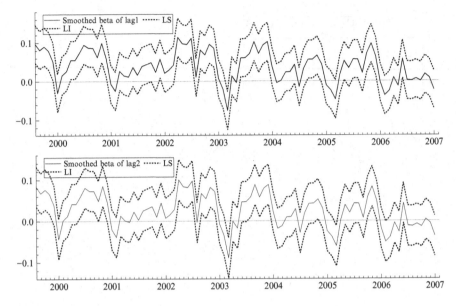

Fig. 5.8 IPC smoothed betas

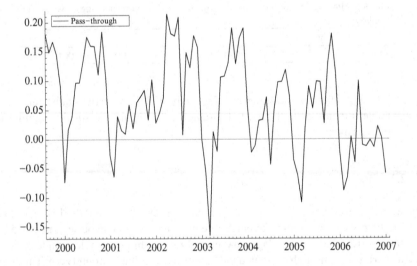

Fig. 5.9 IPC long-run pass-through

Table 5.7 IPC information criteria of unrestricted and restricted models

Criterion	Unrestricted	$\sum_{i=1}^{2} \beta_{i,t} = 0$	$\sum_{i=1}^{2} \beta_{i,t} = 1$
AIC	2.838	2.801	5.291
BIC	3.144	3.051	5.541

5.3 Case III: GDP Benchmarking Estimation and Prediction

5.3.1 Motivation

I close the applications of this book by revisiting the GDP quarterly prediction considered in Pizzinga (2010). Here is the setting. There are two series, a quarterly series of GDP that is subject to measurement error and an annual total series of the same economic variable that is "accurately" recorded. The goal is to produce a quarterly GDP free from those measurement errors, and this is should be accomplished by conveniently using the information from the annual totals. This is in fact a *benchmarking* problem, and its general formulation was examined in a rather comprehensive state space fashion by Durbin and Queenneville (1997) and revisited in Durbin and Koopman (2001), Chap. 3.

5.3.2 Model Setup

Here the focus is on making predictions under this benchmarking framework that will generate quarterly predictions based on GDP free from measurement error and under *consistency* (that is, the estimated quarterly GDP must sum up to the annual GDP totals). For this purpose, Pizzinga (2010) use the restricted Kalman predictor of Sect. 4.3 with an alternative state space form. This representation is an augmented state space model whose augmentations only appear in time period multiples of 4 (four): that is, in these time periods the information from the annual GDP totals is attached to the measurement equation (this makes use of the time- and size-varying flexibility of augmented restricted Kalman filtering). In this sense, the measurements would be Y_t if we "are not" in $4i$ and $(Y_t, X_t)'$ if we "are" in $4i$, where Y_t represents some quarterly GDP, X_t represents the total GDP of some year, and $i = 1, 2, \ldots$. The state vector would be $\alpha_t \equiv (\mu_t, \mu_{t-1}, \mu_{t-2}, \mu_{t-3}, \gamma_t, \gamma_{t-1}, \gamma_{t-2}, \gamma_{t-3}, \varepsilon_t, \varepsilon_{t-1}, \varepsilon_{t-2}, \varepsilon_{t-3}, \xi_t)'$, where μ_t is a local level, γ_t is a dummy seasonal effect, ε_t is a Gaussian white noise irregular component, and ξ_t represents the $AR(1)$ measurement error. In addition, one must set $H_t \equiv 0$, $d_t \equiv 0$ and $c_t \equiv 0$. Finally, verify below the Z_t matrices for this alternative restricted state space form:

$$
Z_t = \begin{cases} [1\,0\,0\,0\,1\,0\,0\,0\,1\,0\,0\,0\,1], & if\ t \neq 4i,\ i = 1, 2, \ldots \\[2mm] \begin{bmatrix} 1\,0\,0\,0\,1\,0\,0\,0\,1\,0\,0\,0\,1 \\ 1\,1\,1\,1\,1\,1\,1\,1\,1\,1\,1\,1\,0 \end{bmatrix}, & if\ t = 4i,\ i = 1, 2, \ldots \end{cases}
$$

The matrices $T_t \equiv T$ and $R_t \equiv R$ and $Q_t \equiv Q$ are obvious and are omitted to save space. Observe that the specification is based purely on the structural modeling

Table 5.8 Results of
benchmarking prediction

Year/quarter	1st	2nd	3rd	4th	Annual total
1995	1,066	1,153	1,129	1,096	4,444
1996	1,030	1,165	1,167	1,139	4,502

framework (see Harvey 1989, Chap. 2) for the quarterly GDP series. Also notice
that the *theoretical* consistency correction acts on a time index multiple of four.
From Theorem 3.1 and from the computational algorithm described at the end of
Sect. 4.3, the *empirical* consistency correction is achieved with in-sample and out-
of-sample periods (the latter would be the prediction) whenever Kalman updating
and smoothing equations actuate on a series extended in the way proposed in (4.10)
(notice that $q_t = X_t$ for every t multiple of 4).

5.3.3 Empirical Results

To illustrate the proposed benchmarking prediction model, Pizzinga (2010) applied
it to the Brazilian GDP series constructed by the methodology proposed in Cerqueira
et al. (2009). This original series was obtained from IBGE (www.ibge.gov.br) and
IPEADATA (www.ipeadata.gov.br). The model estimation was performed using 140
observations ranging from the first quarter of 1960 to the fourth quarter of 1994, a
21.7-s task. In the sequel, a restricted Kalman predictor for the next 2 years using
the annual totals of 1995 and 1996 was used to satisfy the consistency restrictions.
Table 5.8 presents the prediction results. The reader can easily confirm that the
predicted quarterly GDP is consistent with the annual totals.

Chapter 6
Further Extensions

At the end of this book, I list some potential research themes about restricted Kalman filtering. Some of them may already be under investigation.

Firstly, I cite additional *theoretical* points that sound interesting within the theme. These are as follows:

- A study on state observability and parameter identification, which are two important issues to firmly establish the inferential grounds for state space models under linear restrictions.
- A formal investigation into the possible connections between the results in Simon and Chia (2002) and the proofs of restricted Kalman filtering revisited in this book.
- Formal analytical or Monte Carlo investigations into how the presumed additional information due to the use of augmented restricted Kalman filtering translates into improvements for (*quasi*) maximum-likelihood estimators.
- Derivation of results on combining diffuse initialization and linear restrictions under the approach by de Jong and Chu-Chun-Lin (2003) and, consequently, an analysis of how the new assumptions needed are more or less stringent than those considered in Sect. 3.4.

Now, I concentrate on additional *methods*:

- Implementation of an *extended* restricted Kalman filtering to obtain not only nonlinear equality constraints but also *inequality* constraints. In this respect, specific topics of interest would be the investigation of the convergence of this extended approach and how this should be combined with *quasi* maximum-likelihood estimation for fixed parameters.
- Derivation of new tests for coefficient stability under linear restrictions, for which the material from Sect. 3.3 could be of some value.

Lastly, I believe the following *applications* would be relevant to further illustrate and validate already developed methodologies.

A. Pizzinga, *Restricted Kalman Filtering: Theory, Methods, and Application*,
SpringerBriefs in Statistics 12, DOI 10.1007/978-1-4614-4738-2_6,
© Springer Science+Business Media New York 2012

- Estimation of a dynamic factor model with an exact *smoothing transition* coefficient under the same linear and interpretable portfolio restrictions and also under some linear restrictions regarding *leverage/hedge*.
- Formulation and estimation of *multivariate* benchmarking models aimed at prediction.

References

ANDERSON, B. D. O. and MOORE, J. B. (1979). *Optimal Filtering*. Prentice Hall.

ANSLEY. C. F. and KOHN, R. (1985). "Estimation, filtering and smoothing in state space models with incompletely specified initial conditions". *Annals of Statistics*, Vol. 13, No. 4, pp. 1286–1316.

BROCKWELL, P. J. and DAVIS, R. A. (1991). *Time Series: Theory and Methods*. 2nd edition. Springer.

BROCKWELL, P. J. and DAVIS, R. A. (2003). *Introduction to Time Series and Forecasting*. 2nd edition. Springer Texts in Statistics.

BROWN, R., DURBIN, J. and EVANS, J. (1975). "Techniques for testing the constancy of regression relationships Over Time". *Journal of the Royal Statistical Society B*, 37, pp. 149–172.

CAMPA, J.M. and GOLDBERG, L.S. (1995). *Exchange Rate Pass-through into Import Prices*. Federal Reserve Bank of New York.

CARHART, M. M. (1997). "On persistence in mutual fund performance". *Journal of Finance*, 52, pp. 57–82.

CERQUEIRA, L. F., PIZZINGA, A., and FERNANDES, C. (2009). "Methodological procedure for estimating Brazilian quarterly GDP series". *International Advances in Economic Research*, 15, pp. 102–114.

CHUNG, K. L. (2001). *A Course in Probability Theory*. 3rd edition. Academic Press.

DAVIDSON, R. and MACKINNON, J. G. (1993). *Estimation and Inference in Econometrics*. Oxford University Press.

DE ROON, F. A., NIJMAN, T. E. and TER HORST, J. R. (2004). "Evaluating style analysis". *Journal of Empirical Finance*, Volume 11, Issue 1, pp. 29–53.

DE JONG, P. and ZEHNWIRTH, B. (1983). "Claims reserving state-space models and the Kalman filter". *Journal of the Institute of Actuaries*, 110, pp. 157–181.

DE JONG, P. (1988). "The likelihood for a state space model". *Biometrika*, 75, 1, pp. 165–169.

DE JONG, P. (1989). "Smoothing and Interpolation With the State-Space Models". *Journal of the American Statistical Association*, 84, pp. 1085–1088.

DE JONG, P. (1991). "The diffuse Kalman filter". *Annals of Statistics*, Vol. 19, No. 2, pp. 1073–1083.

DE JONG, P. and CHU-CHUN-LIN, S. (2003). "Smoothing with an unknown initial condition". *Journal of Time Series Analysis*, Vol. 24, No. 2, pp. 141–148.

DOORNICK, J. A. (2001). *Ox 3.0: An Object-Oriented Matrix Programming Language*. Timberlake Consultants.

DORAN, H. (1992). "Constraining Kalman filter and smoothing estimates to satisfy time-varying restrictions". *Review of Economics and Statistics*, 74, pp. 568–572.

A. Pizzinga, *Restricted Kalman Filtering: Theory, Methods, and Application*,
SpringerBriefs in Statistics 12, DOI 10.1007/978-1-4614-4738-2,
© Springer Science+Business Media New York 2012

DORAN, H. (1996). "Estimation under exact linear time-varying constraints, with applications to population projections". *Journal of Forecasting*, 15, pp. 527–541.

DORAN, H. and RAMBALDI, A. (1997). "Applying linear time-varying constraints to econometric models: with an application to demand systems". *Journal of Econometrics*, 79, pp. 83–95.

DURBIN, J. and KOOPMAN, S. J. (2001). *Time Series Analysis by State Space Methods*. Oxford Statistical Science Series.

DURBIN, J. and QUEENNEVILLE, B. (1997). "Benchmarking by state space models". *International Statistical Review*, 65, pp. 21–48.

ELTON, E. J., GRUBER, M. J., BROWN, S. J. and GOETZMANN, W. (2006). *Modern Portfolio Theory and Investment Analysis*. 7th edition. John Wiley & Sons.

ENDERS, W. (2004). *Applied Econometric Time Series*. 2nd edition. John Wiley & Sons.

FRAGA, A., GOLDFAJN, I. and MINELLA, A. (2003) "Inflation Targeting in Emerging Market Economies". *NBER Working Paper*, 10.019.

GEETER, J., BRUSSEL, H. and SCHUTTER, J. (1997). "A smoothly constrained Kalman filter". *IEEE Transactions on Pattern Analysis and Machine Intelligence*, 19, 10, pp. 1171–1177.

GREENE, W. H. (2003), *Econometric Analysis*. 5th edition. Prentice Hall.

HAMILTON, J. D. (1994). Time Series Analysis. Princeton University Press.

HARVEY, A. C. (1981). *The Econometric Analysis of Time Series*. Philip Allan Publishers.

HARVEY, A. C. (1989). *Forecasting, Structural Time Series Models and The Kalman Filter*. Cambridge University Press.

HARVEY, A. C. (1993). *Time Series Models*. 2nd edition. Harvester Wheatsheaf.

JENSEN, M. C. (1968). "The performance of mutual funds in the period 1945–1964". *Journal of Finance*, Vol. 23, No. 2, pp. 389–416.

JOHNSTON, J. and DiNARDO, J. (1997). *Econometric Methods*. 4th edition. McGraw-Hill.

JULIER, S. and UHLMANN, J. (2004). "Unscented filtering and nonlinear estimation". *Proceedings of the IEEE*, 92, pp. 401–422.

JULIER, S. and LAVIOLA, J. (2007). "Kalman filtering with nonlinear equality con- straints". *IEEE Transactions on Signal Processing*, 55, 6, pp. 2774–2784.

KO, S. and BITMEAD, R. R. (2007). "State estimation for linear systems with state equality constraints". *Automatica*, 43, 8, pp. 1363–1368.

KOOP, G., LEON-GONZALES, R. and STRATCHAN, R. (2010). "Dynamic probabilities of restrictions in state space models: an application to the Phillips curve". *Journal of Business & Economic Statistics*, Vol. 28, No. 3, pp. 370–379.

KOOPMAN, S. J. (1997). "Exact initial Kalman filtering and smoothing for nonstationary time series models". *Journal of the American Statistical Association*, 92, pp. 1630–1638.

KOOPMAN, S. J., SHEPHARD, N. and DOORNIK, J. A. (2002). "SsfPack 3.0 beta02: statistical algorithms for models in state space". *Unpublished paper*. Department of Econometrics, Free University, Amsterdam.

KOOPMAN, S. J. and DURBIN, J. (2003). "Filtering and smoothing of state vector for diffuse state-space models". *Journal of Time Series Analysis*, Vol. 24, No. 1, pp. 85–98.

KUBRUSLY, C. S. (2001). *Elements of Operator Theory*. Birkhäuser.

LEYBOURNE, S. (1993). "Estimation and testing of time-varying coefficient regression models in the presence of linear restrictions". *Journal of Forecasting*, 12, pp. 49–62.

MASSICOTTE, D., MORAWSKI, R. Z. and BARWICZ, A. (1995). "Incorporation of a positivity constraint into a Kalman-filter-based algorithm for correction of spectrometric data". *IEEE Transactions on Instrumentation and Measurement*, 44, 1, pp. 2–7.

MENON, J. (1996). "Exchange rate pass-through". *Journal of Economic Surveys*, Vol. 8, No. 2, pp. 197–231.

PAGAN, A. (1980). "Some identification and estimation results for regression models with stochastically varying coefficients". *Journal of Econometrics*, 13, pp. 341–363.

PANDHER, G. S. (2002). "Forecasting multivariate time series with linear restrictions using constrained structural state-space models". *Journal of Forecasting*, 21, pp. 281–300.

PANDHER, G. S. (2007). "Modelling & controlling monetary and economic identities with constrained state space models". *International Statistical Review*, Vol. 75, No. 2, pp. 150–169.

PARSLEY, D. (1995). "Anticipated future shocks and exchange rate pass-through in the presence of reputation". *International Review of Economics* Vol. 4, No. 2, pp. 99–103.

PIZZINGA, A. (2009). "Further investigation into restricted Kalman filtering". *Statistics & Probability Letters*, 79, pp. 264–269.

PIZZINGA, A. (2010). "Constrained Kalman filtering: additional results". *International Statistical Review*, Vol. 78, No. 2, pp. 189–208.

PIZZINGA, A. (2012)"Diffuse restricted Kalman filtering". *Communications in Statistics: Theory and Methods* (in press).

PIZZINGA, A., RUGGERI, E. and GUEDES, Q. (2005). "Relatório Técnico Estatístico: Projeto Hooke (in Portuguese)". *Technical report*. DCT.T/Furnas Centrais Elétricas S.A.

PIZZINGA, A. and FERNANDES, C. (2006). "State space models for dynamic style analysis of portfolios". *Brazilian Review of Econometrics*, Vol. 26, 1, pp. 31–66.

PIZZINGA, A., FERNANDES, C. and CONTRERAS, S. (2008a). "Restricted Kalman filtering revisited". *Journal of Econometrics*, 144, 2, pp. 428–429.

PIZZINGA, A., VEREDA, L., ATHERINO, R. and FERNANDES, C. (2008b). "Semi-strong dynamic style analysis with time-varying selectivity measurement: applications to Brazilian exchange rate funds". *Applied Stochastic Models in Business and Industry*, Vol. 24, 1, pp. 3–12.

PIZZINGA, A., VEREDA, L. and FERNANDES, C. (2011). "A dynamic style analysis of exchange rate funds: the case of Brazil at the 2002 election". *Advances and Applications in Statistical Sciences*. Vol. 6, Issue 2, pp. 111–135.

SHARPE, W. F. (1988). "Determining a fund's effective asset mix". *Investment Management Review*, pp. 59–69.

SHARPE, W. F. (1992). "Asset allocation: management style and performance measurement". *Journal of Porfolio Management*, Winter, pp. 7–19.

SHUMWAY, R. H. and STOFFER, D. S. (2006). *Time Series Analysis and Its Applications (With R Examples)*. Springer.

SIMON, D. (2009). "Kalman Filtering with State Constraints: a Survey of Linear and Nonlinear Algorithms". *IET Control Theory & Applications*. (in press)

SIMON, D. and CHIA, T. (2002). "Kalman filtering with state equality constraints". *IEEE Transactions on Aerospace and Electronic Systems*, 38, 1, pp. 128–136.

SIMON, D. and SIMON, D. L. (2004). "Aircraft turbofan engine health estimation using constrained Kalman filtering". *Journal of Engineering for Gas Turbines and Power*, 126, pp. 1–6.

SOUZA, R. M., MACIEL, L. and PIZZINGA, A. (2011). "Using a restricted Kalman filtering approach for the estimation of a dynamic exchange-rate pass-through". In Gomez, J. M. (ed.) *Kalman Filtering*, Chap. 9, pp. 255–268. Nova Publishers.

SWINKELS, L. and VAN DER SLUIS, P. J. (2006). "Return-based style analysis with time-varying exposures". *European Journal of Finance*, Vol. 12, pp. 529–552.

TANIZAKI, H. (1996). *Nonlinear Filters*. 2nd edition. Springer.

TAYLOR, J. (2000) "Low inflation, pass-through and the pricing power of firms". *European Economic Review*, 44, pp. 1389–1408.

TEIXEIRA, B. O. S., CHANDRASEKAR, J., TORRES, L. A. B., AGUIRRE, L. A. and BERNSTEIN, D. S. (2009). "State estimation for linear and nonlinear equality-constrained systems". *International Journal of Control*, 82, 5, pp. 918–936.

WEST, M. and HARRISON, J. (1997). *Bayesian Forecasting and Dynamic Models*. 2nd edition. Springer.